Complex Space Source Theory of Spatially Localized Electromagnetic Waves

Mario Boella Series on Electromagnetism in Information and Communication

Piergiorgio L. E. Uslenghi, PhD – Series Editor

Mario Boella series offers textbooks and monographs in all areas of radio science, with a special emphasis on the applications of electromagnetism to information and communication technologies. The series is scientifically and financially sponsored by the Istituto Superiore Mario Boella affiliated with the Politecnico di Torino, Italy, and is scientifically co-sponsored by the International Union of Radio Science (URSI). It is named to honor the memory of Professor Mario Boella of the Politecnico di Torino, who was a pioneer in the development of electronics and telecommunications in Italy for half a century, and a vice president of URSI from 1966 to 1969.

Published Titles in the Series

Introduction to Wave Phenomena
by Akira Hirose and Karl Lonngren

Scattering of Waves by Wedges and Cones with Impedance Boundary Conditions
by Mikhail Lyalinov and Ning Yan Zhu

Complex Space Source Theory of Spatially Localized Electromagnetic Waves
by S. R. Seshadri

Forthcoming Titles

The Wiener-Hopf Method in Electromagnetics
by Vito Daniele and Rodolfo Zich (2014)

Higher Order Numerical Solution Techniques in Electromagnetics
by Roberto Graglia and Andrew Peterson (2014)

Slotted Waveguide Array Antennas
by Sembiam Rengarajan and Lars Josefsson (2015)

Wideband and Time Varying Antenna Arrays
by Randy Haupt (2015)

Complex Space Source Theory of Spatially Localized Electromagnetic Waves

ISMB Series

S. R. Seshadri

SciTech
PUBLISHING
an imprint of the IET

Edison, NJ
scitechpub.com

Published by SciTech Publishing, an imprint of the IET.
www.scitechpub.com
www.theiet.org

10 9 8 7 6 5 4 3 2 1

ISBN 978-1-61353-193-8 (hardback)
ISBN 978-1-61353-194-5 (PDF)

Typeset in India by MPS Limited
Printed in the US by Integrated Books International
Printed in the UK by Berforts Information Press Ltd.

To the memory
of
my parents

Contents

Preface   xiii
Foreword   xvii

1   Fundamental Gaussian beam   **1**
  1   Vector potential   1
  2   Electromagnetic fields   3
  3   Radiation intensity   4
  4   Radiative and reactive powers   6
  5   Beam spreading on propagation   7
  6   Magnetic current density   8
  7   Some applications and limitations   9
       References   10

2   Fundamental Gaussian wave   **13**
  1   Exact vector potential   13
  2   Exact electromagnetic fields   15
  3   Radiation intensity   17
  4   Radiative and reactive powers   19
  5   Gaussian beam beyond the paraxial approximation   21
       References   23

3   Origin of point current source in complex space   **25**
  1   Scalar Gaussian beam   26
  2   Field of a point source   28

3	Extensions	29
4	Exact solution	30
	References	30
	Gaussian beam as a bundle of complex rays	31

4 Basic full Gaussian wave — **35**

1	Point source in complex space	35
2	Electromagnetic fields	37
3	Radiation intensity	39
4	Radiative and reactive powers	40
5	General remarks	42
	References	43

5 Complex source point theory — **45**

1	Derivation of complex space source	45
2	Asymptotic field	47
3	Analytic continuation	48
4	Limiting absorption	50
	References	53

6 Extended full Gaussian wave — **55**

1	Current source of finite extent in complex space	55
2	Time-averaged power	58
3	Radiation intensity	61
4	Radiative and reactive powers	63
	References	66

7 Cylindrically symmetric transverse magnetic full Gaussian wave — **67**

1	Cylindrically symmetric transverse magnetic beam	67
2	Current source in complex space	72
3	Cylindrically symmetric TM full wave	74
4	Real power	75
5	Reactive power	78
6	Radiation intensity distribution	80
	References	82

8 Two higher-order full Gaussian waves   **85**

   1   Higher-order hollow Gaussian wave   85

      1.1 Paraxial beam   85

      1.2 Complex space source   87

      1.3 Hollow Gaussian wave   88

   2   cosh-Gauss wave   90

      2.1 cosh-Gauss beam   90

      2.2 Complex space source   92

      2.3 cosh-Gauss wave   93

   References   94

9 Basic full complex-argument Laguerre–Gauss wave   **97**

   1   Complex-argument Laguerre–Gauss beam   97

      1.1 Paraxial beam   97

      1.2 Time-averaged power   100

   2   Complex space source   101

   3   Complex-argument Laguerre–Gauss wave   103

   4   Real and reactive powers   104

   References   109

10 Basic full real-argument Laguerre–Gauss wave   **111**

   1   Real-argument Laguerre–Gauss beam   111

      1.1 Paraxial beam   111

      1.2 Complex power   114

   2   Real-argument Laguerre–Gauss wave   116

   3   Real and reactive powers   118

   References   122

11 Basic full complex-argument Hermite–Gauss wave   **123**

   1   Complex-argument Hermite–Gauss beam   123

      1.1 Paraxial beam   123

      1.2 Time-averaged power   126

   2   Complex-argument Hermite–Gauss wave   128

   3   Real and reactive powers   130

   References   135

12 Basic full real-argument Hermite–Gauss wave   137

    1   Real-argument Hermite–Gauss beam   137

       1.1 Paraxial beam   137

       1.2 Time-averaged power   139

    2   Real-argument Hermite–Gauss wave   140

    3   Real and reactive powers   142

    Appendix AA: Evaluation of the integral $I_m(x,z)$ 147

    Appendix AB: Complex space source   148

    References   149

13 Basic full modified Bessel–Gauss wave   151

    1   Modified Bessel–Gauss beam   151

       1.1 Paraxial beam   151

       1.2 Time-averaged power   153

    2   Modified Bessel–Gauss wave   155

    3   Real and reactive powers   156

    References   158

14 Partially coherent and partially incoherent full Gaussian wave   161

    1   Extended full-wave generalization of the paraxial beam   162

    2   Cross-spectral density   165

    3   Radiation intensity for the partially coherent source   169

    4   Time-averaged power for the partially coherent source   173

    5   Radiation intensity for the partially incoherent source   175

    6   Time-averaged power for the partially incoherent source   178

    7   General remarks   178

    References   180

15 Airy beams and waves   181

    1   Fundamental Airy beam   181

    2   Modified fundamental Airy beam   184

    3   Fundamental Airy wave   187

    2   Basic full modified Airy wave   190

    5   Remarks   197

    References   197

Appendix A   Green's function for the Helmholtz equation   199

A.1   Three-dimensional scalar Green's function   199

A.2   Fourier transform of scalar Green's function   201

A.3   Bessel transform of scalar Green's function   203

Reference   204

Appendix B   An integral   205

Appendix C   Green's function for the paraxial equation   207

C.1   Paraxial approximation   207

C.2   Green's function $G_p(x, y, z)$ 208

C.3   Fourier transform $\overline{G}_p(p_x, p_y, z)$ 209

C.4   Bessel transform $\overline{G}_p(\eta, z)$ 209

References   211

Appendix D   Electromagnetic fields   213

D.1   Poynting vector and generated power   213

D.2   Vector potentials   216

References   217

Appendix E   Airy integral   219

References   220

Index   221

Preface

Electromagnetic waves are governed by Maxwell's equations. Exact solutions of Maxwell's equations are obtained from a single component of the magnetic vector potential and a single component of the electric vector potential, both in the same direction. These two scalar wave functions satisfy the Helmholtz equation. This monograph presents electromagnetic waves which are exact solutions of the governing Helmholtz equation and which are spatially localized around a specified direction. The spatial localization is achieved in a particular manner. The Helmholtz equation is approximated for a wave that has wavevectors in a narrow range of directions about the propagation direction, usually taken as the z axis. In this approximation, the wave function is separated out into a rapidly varying plane-wave phase appropriate for the propagation direction and a slowly varying amplitude that satisfies the paraxial wave equation. The solution of the paraxial wave equation yields the paraxial beams. The approximate paraxial beam solutions are generalized to yield the full waves that are exact solutions of the Helmholtz equation. Generally, "beam" is used to characterize the spatially localized fields derived from the paraxial wave equation and "full wave" or "wave" is used to describe the spatially localized fields determined from the Helmholtz equation.

The paraxial beam approximation has many drawbacks. This approximation is not valid when the beam waist is comparable to or less than the wavelength. The paraxial beam approximation usually does not take into account the polarization properties of the electromagnetic fields. Paraxial beam approximation is not suitable for the treatment of the near fields. In some cases, the paraxial beam solutions cannot be physically realized. Thus, there is a need to obtain exact full-wave solutions that reduce to the paraxial beams in the appropriate limit. There is a whole class of full-wave solutions that reduce to the same paraxial beam solution in the proper limit. This monograph treats the exact full-wave generalizations of many of the basic types of paraxial beam solutions.

There are two major steps in obtaining the full wave generalizations of the paraxial beam approximations. First is the systematic derivation of the appropriate virtual source in the complex space that produces the required full wave from the paraxial beam solution. At the beginning, a point source in the complex space was postulated to obtain the Gaussian beam in the real physical space. Subsequently, a few other sources such as a combination of point sources or higher-order point sources were postulated to obtain particular paraxial beams. Now the required source in the complex space is derived starting from the specific paraxial beam solution. The second step in the theory is the determination of the actual secondary source in the physical space that is equivalent to the virtual source in the complex space. From the secondary source, the dynamics of all the full waves such as the input impedance of the source and the radiation intensity distribution are obtained. These new theoretical developments were published in recent years in the *Journal of the Optical Society of America* A.

This research monograph is new in a number of ways, with many new topics covered. New analytical techniques are introduced, and for a scholarly monograph, the method of presentation is new in the sense that all analytical steps are carefully explained. The treatment is presented in a self-contained manner that will enable second-year graduate students in applied physics, electrical engineering, physics, and applied mathematics to learn the subject and proceed to perform research in the area.

The needed discussion of paraxial beams is included in the treatment of the complex space source theory of spatially localized electromagnetic waves, but the paraxial beams are not treated *per se*. The fundamental Gaussian beam, the cylindrically symmetric transverse magnetic Gaussian beam, the higher-order hollow Gaussian beam, the cosh-Gauss beam, the complex-argument and the real-argument Laguerre–Gauss beams, the complex-argument and the real-argument Hermite–Gauss beams, and the modified Bessel–Gauss beam are treated. Material normally present in the treatment of traditional paraxial beams and their applications is not duplicated.

The amplitude of the electromagnetic field varies in a random manner on a time scale of several hundred times the wave period. The Poynting vector in the propagation direction averaged over a wave period is expressed in terms of the cross-spectral density of the fluctuating vector potential across the input plane. The input value of the cross-spectral density takes account of the correlations of the electromagnetic fields between two points in the source plane and fully governs the propagation characteristics such as the radiation intensity distribution and the radiated power. The partially coherent and partially incoherent electromagnetic fields are rarely covered in the traditional treatment of paraxial beams. In this monograph, not only are the partially coherent and partially incoherent electromagnetic waves fully covered, but the treatment extends beyond the paraxial approximation to include the full waves governed by Maxwell's equations.

In general, the Gaussian beam is the foundation for the paraxial beams and full waves. The input field of the Gaussian beam has a waist, and the Rayleigh distance is defined in terms of the waist. For the full waves based on the Gaussian beam, the source location in the complex space is known in terms of the Rayleigh distance. There is a new class of paraxial beams based on the Airy function. The fundamental Airy beam is not physically realizable. The Airy function with a superimposed exponential function that causes the amplitude to decrease for negative values of the argument leads to a "finite-energy" fundamental Airy beam that is physically realizable. The full-wave generalization of the modified fundamental Airy beam is constructed in terms of higher-order point sources in the complex space. But the source location is not known *a priori*. From the wavenumber representation of the modified fundamental Airy beam, the source location is systematically deduced. Based on the source location, an equivalent beam waist is introduced. The propagation characteristics of the full-wave generalization of the modified fundamental Airy beam expressed in terms of the equivalent waist are identical to the full-wave generalization of the fundamental

Gaussian beam expressed in terms of the actual waist. For the basic full modified funda-mental Airy wave, the source is an infinite series of point sources with an increasing order starting from zero. A brief treatment of the newly developing area of Airy beams and waves is given in the last chapter. The complex space source theory of the Airy waves is carefully explained.

The approximate paraxial beam solutions and the exact full-wave generalizations are developed by the use of Fourier and Bessel transform techniques. An effort was made to integrate the complex space source theory of electromagnetic waves as a branch of Fourier optics.

A note on the method of numbering of the equations, figures, and sections is useful. When reference is made to an equation as (4.5), the first number indicates the chapter to which the equation pertains and the second gives the number of the equation. When a reference is made to an equation in the same chapter, the first number is omitted. If reference is made to a figure as 14.5, the first number gives the chapter number and the second the figure number. When reference is made to a section as 2.3, the first number is the chapter number and the second number gives the section number. When reference is made to a figure or a section in the same chapter, the first number is omitted. Appendices appear at the end of the monograph and at the end of Chapter 12. The former are identified by the letters A, B, C, D, and E, and the latter by the letters AA and AB. When reference is made to an equation as (B4), the first letter indicates Appendix B at the end of the monograph and the letter and number indicate the equation number. When reference is made to an equation as (AB6), the two letters indicate Appendix AB at the end of Chapter 12, and the two letters and number denote the equation number.

During the time of preparation of this monograph, I have read with advantage the pub-lications of several previous authors whose works are listed at the end of chapters and appendices.

I am indebted to Piergiorgio L. E. Uslenghi, Series Editor, and Dudley R. Kay, Publisher, for their valuable suggestions for the revision of the manuscript, and to Rachel Williams of the IET and Vijay Ramalingam of MPS Limited for their helpful guidance in the production phase of the monograph. The monograph is scientifically and financially sponsored by the Istituto Superiore Mario Boella (ISMB) and scientifically cosponsored by the Union Radio-Scientifique Internationale (URSI). To these two institutions, ISMB and URSI, I am grateful.

I dedicate this monograph respectfully to the memory of my parents, M. S. Srinivasan and Doriammal.

Livermore, California   S. R. Seshadri
May 2013

Foreword

Over the past few decades, there has been a resurgent interest in directed propagation of electromagnetic energy for at least four reasons: first, the development of laser technology has made it necessary to understand the localized propagation of optical energy in terms of Gaussian beams; second, concerted attempts have been made to create microwave sources for the generation of high-intensity narrow beams for military purposes; third, studies have been conducted on harnessing solar energy on satellite stations and transferring that energy to ground stations at microwave frequencies in order to penetrate cloud cover; fourth, there have been extensive studies of electromagnetic beams propagation through penetrable tissues in biomedicine.

From a theoretical viewpoint, all the above applications are rooted in understanding the characteristics and limitations of focused electromagnetic wave propagation through a variety of media, on the basis of Maxwell's equations. The present monograph by S. R. Seshadri addresses some of the fundamental issues in this very important field. It provides a solid theoretical basis for graduate students and researchers to understand the fundamental concepts behind directed propagation of electromagnetic energy, and gives them the basic background on which to conduct additional, pioneering research.

Piergiorgio L. E. Uslenghi – Series Editor
Chicago, August 2013

CHAPTER 1

Fundamental Gaussian beam

A plane wave has a unique propagation direction and is not physically realizable since infinite energy is required for its launching. Nearly plane waves or beams are formed by a group of plane waves having a narrow range of propagation directions about a specified direction. A general electromagnetic field is constructed from a single component of magnetic vector potential and a single component of electric vector potential, both in the same direction. The vector potential associated with an electromagnetic beam is separated into a rapidly varying phase and a slowly varying amplitude. The slowly varying amplitude satisfies the paraxial wave equation. For an input distribution having a simple Gaussian profile with circular cross section, the paraxial wave equation is solved to obtain the vector potential. For the fundamental electromagnetic Gaussian beam, the fields are evaluated and the characteristics of the radiation intensity distribution are described. The outward propagations in the $+z$ direction in the space $0 < z < \infty$ and in the $-z$ direction in the space $-\infty < z < 0$ are considered. The secondary source is concentrated on the boundary plane $z = 0$. The source current density is obtained and the complex power is determined. The time average of the real power is equal to the time-averaged radiative power. The reactive power of the paraxial beam is found to vanish. Treatments of the fundamental Gaussian beam together with additional analytical developments are available in the literature [1–8].

1 Vector potential

The outward propagations in the $+z$ direction in the space $0 < z < \infty$ and in the $-z$ direction in the space $-\infty < z < 0$ are considered. The time dependence is harmonic of the form: $\exp(-i\omega t)$, where $\omega/2\pi$ is the wave frequency. The plane $z = 0$, which forms the boundary between the two half spaces, is the secondary source plane (see page 239 in [7]). The secondary source is an infinitesimally thin sheet of electric current. The secondary source current density vanishes for $z \neq 0$ and is infinite for $z = 0$. The electric current is in the x direction and excites the x component of the magnetic vector potential. This potential is used to construct the electromagnetic fields of the linearly polarized fundamental Gaussian light beam [see Eqs. (D25) and (D30)–(D33)]. For generating the linearly polarized fundamental Gaussian light beam, the paraxial approximation of the x component of the magnetic vector potential on the input plane is assumed as

$$A_{x0}^{\pm}(x, y, 0) = \frac{N}{ik} \exp\left(-\frac{x^2 + y^2}{w_0^2}\right) \tag{1}$$

where k is the wavenumber and w_0 is the waist size of the beam at the input plane $z = 0$. The subscript 0 denotes paraxial. The sign $\pm$, the superscript $\pm$, and the subscript $\pm$ indicate that the propagation is in the $\pm z$ direction. The total time-averaged power $P_0^{\pm}$ transported by the paraxial beam in the $\pm z$ direction is the same. The normalization constant N is chosen such that $P_0^{\pm} = 1$ W. This requirement yields the normalization constant as

$$N = (4/c\pi w_0^2)^{1/2} \tag{2}$$

where c is the electromagnetic wave velocity in free space. The two-dimensional Fourier transform of Eq. (1) is evaluated by the use of Eqs. (A18), (B1), and (B6) as

$$\overline{A}_{x0}^{\pm}(p_x, p_y, 0) = \frac{N}{ik} \pi w_0^2 \exp\left[-\pi^2 w_0^2 \left(p_x^2 + p_y^2\right)\right] \tag{3}$$

The rapidly varying phase $\exp(\pm ikz)$ for the propagation in the $\pm z$ direction is separated out from the paraxial approximation of the x component of the magnetic vector potential as

$$A_{x0}^{\pm}(x, y, z) = \exp(\pm ikz) a_{x0}^{\pm}(x, y, z) \tag{4}$$

For plane waves, $a_{x0}^{\pm}(x, y, z)$ is a constant. For nearly plane waves or beams, $a_{x0}^{\pm}(x, y, z)$ is a slowly varying function of its arguments. $A_{x0}^{\pm}(x, y, z)$ satisfies the Helmholtz equation [see Eqs. (D25) and (C1)]. When $A_{x0}^{\pm}(x, y, z)$ is substituted into the Helmholtz equation, in the paraxial approximation, $a_{x0}^{\pm}(x, y, z)$ is found to satisfy the paraxial equation:

$$\left(\frac{\partial^2}{\partial x^2} + \frac{\partial^2}{\partial y^2} \pm 2ik \frac{\partial}{\partial z}\right) a_{x0}^{\pm}(x, y, z) = 0 \tag{5}$$

See Appendix C. The two-dimensional Fourier integral representation of $a_{x0}^{\pm}(x, y, z)$ as given by Eq. (A17) is used. Then $\overline{a}_{x0}^{\pm}(p_x, p_y, z)$, the two-dimensional Fourier transform of $a_{x0}^{\pm}(x, y, z)$, is found to satisfy a one-dimensional differential equation in z such as that given by Eq. (C10). The solution of the differential equation is expressed as

$$\overline{a}_{x0}^{\pm}(p_x, p_y, z) = \overline{a}_{x0}^{\pm}(p_x, p_y, 0) \exp\left[-\pi^2 w_0^2 \left(p_x^2 + p_y^2\right) \frac{i|z|}{b}\right] \tag{6}$$

where $b = \frac{1}{2} k w_0^2$ is the Rayleigh distance. In the paraxial approximation, for the entire range of the wavenumbers $2\pi p_x$ and $2\pi p_y$ in the transverse directions, the longitudinal wavenumber in the propagation direction $(\pm z)$ is found from Eq. (6) to be real. Therefore, in the paraxial approximation, there are no evanescent waves.

From Eqs. (3) and (4), it is found that

$$\overline{a}_{x0}^{\pm}(p_x, p_y, 0) = \frac{N}{ik} \pi w_0^2 \exp\left[-\pi^2 w_0^2 \left(p_x^2 + p_y^2\right)\right] \tag{7}$$

The substitution of Eq. (7) into Eq. (6) leads to $\bar{a}_{x0}^{\pm}(p_x, p_y, z)$. The slowly varying amplitude of the paraxial beam is determined by the inverse Fourier transformation of $\bar{a}_{x0}^{\pm}(p_x, p_y, z)$ as

$$a_{x0}^{\pm}(x, y, z) = \frac{N}{ik} \pi w_0^2 \int_{-\infty}^{\infty} \int_{-\infty}^{\infty} dp_x dp_y \exp[-i2\pi(p_x x + p_y y)]$$

$$\times \exp\left[-\frac{\pi^2 w_0^2 (p_x^2 + p_y^2)}{q_{\pm}^2}\right] \tag{8}$$

where

$$q_{\pm} = \left(1 \pm \frac{iz}{b}\right)^{-1/2} \tag{9}$$

For the position coordinates in the physical space $|z| > 0$, $1/q_{\pm}^2 \neq 0$, and the integrals in Eq. (8) are evaluated by the use of Eqs. (B1) and (B6) as

$$a_{x0}^{\pm}(x, y, z) = \frac{N}{ik} q_{\pm}^2 \exp\left[-\frac{q_{\pm}^2 (x^2 + y^2)}{w_0^2}\right] \tag{10}$$

From Eqs. (4) and (10), $A_{x0}^{\pm}(x, y, z)$ is obtained as

$$A_{x0}^{\pm}(x, y, z) = \exp(\pm ikz) \frac{N}{ik} q_{\pm}^2 \exp\left[-\frac{q_{\pm}^2 (x^2 + y^2)}{w_0^2}\right] \tag{11}$$

2 Electromagnetic fields

The electromagnetic fields of the paraxial beam are determined by substituting $A_{x0}^{\pm}(x, y, z)$ into Eqs. (D30)–(D33). The transverse derivatives $\partial/\partial x$ and $\partial/\partial y$ operating on $A_{x0}^{\pm}(x, y, z)$ act only on the slowly varying amplitude and are equivalent to the introduction of the factor $1/w_0$. The longitudinal derivative operating on the rapidly varying phase of $A_{x0}^{\pm}(x, y, z)$ introduces the factor k, but acting on the slowly varying amplitude introduces the factor $1/b = 2/kw_0^2$. Thus, with reference to the term obtained by $\partial/\partial z$ acting on the rapidly varying phase of $A_{x0}^{\pm}(x, y, z)$, $\partial/\partial x$ and $\partial/\partial y$ operating on $A_{x0}^{\pm}(x, y, z)$ make the resultant quantity one order of magnitude in $1/kw_0$ smaller, and $\partial/\partial z$ acting on the slowly varying amplitude of $A_{x0}^{\pm}(x, y, z)$ makes the resultant quantity two orders of magnitude in $1/kw_0$ smaller. Consequently, the leading term in $E_{x0}^{\pm}$ given by Eq. (D30) is found as $E_{x0}^{\pm} = ikA_{x0}^{\pm}$; the neglected term is two orders of magnitude in $1/kw_0$ smaller than the leading term. Similarly, $E_{z0}^{\pm}$ is two orders of magnitude smaller than the leading term in $E_{x0}^{\pm}$. The leading term in $E_{z0}^{\pm}$ is one order of magnitude smaller than the leading term in $E_{x0}^{\pm}$. The leading term in $H_{y0}^{\pm}$ is given by $\pm ikA_{x0}^{\pm}$ and is of the same order of magnitude as the leading term in $E_{x0}^{\pm}$. Finally, $H_{z0}^{\pm}$ is an order of magnitude smaller than the leading term in $E_{x0}^{\pm}$. Hence, the electromagnetic fields associated with the paraxial beam are given by

$$E_{x0}^{\pm}(x, y, z) = \pm H_{y0}^{\pm}(x, y, z) = ikA_{x0}^{\pm}(x, y, z) \tag{12}$$

We find that $H_{x0}^{\pm} \equiv 0$, $E_{z0}^{\pm}$ and $H_{z0}^{\pm}$ are one order of magnitude smaller, and $E_{y0}^{\pm}$ is two orders of magnitude smaller than that given by Eq. (12). Therefore, the fundamental Gaussian light beam generated by the vector potential $A_{x0}^{\pm}(x, y, z)$ is linearly polarized with the electric field in the x direction and with the magnetic field in the y direction. Thus, xz and yz are the E-plane and the H-plane, respectively.

The time-averaged power flow per unit area in the $\pm z$ direction is found from Eq. (D10) as

$$\pm S_{z0}^{\pm}(x, y, z) = \pm \frac{c}{2} \text{Re}[E_{x0}^{\pm}(x, y, z) H_{y0}^{\pm*}(x, y, z)] \tag{13}$$

where Re stands for the real part and the asterisk denotes complex conjugation. The use of Eqs. (11) and (12) in Eq. (13) yields

$$\pm S_{z0}^{\pm}(x, y, z) = \frac{cN^2}{2(1 + z^2/b^2)} \exp\left[-\frac{2(x^2 + y^2)}{w_0^2(1 + z^2/b^2)}\right] \tag{14}$$

The time-averaged power $P_0^{\pm}$ transported by the paraxial Gaussian beam in the $\pm z$ direction is determined by integrating $[\pm S_{z0}^{\pm}(x, y, z)]$ with respect to x and y across the entire transverse plane, with the result that

$$P_0^{\pm} = \int_{-\infty}^{\infty} \int_{-\infty}^{\infty} dx dy [\pm S_{z0}^{\pm}(x, y, z)] = \frac{c\pi w_0^2 N^2}{4} = 1 \tag{15}$$

The integrals occurring in Eq. (15) are evaluated by the use of Eqs. (B1) and (B6). As indicated previously, the choice of N given by Eq. (2) leads to the total power transported by the paraxial beam in the $\pm z$ direction, equal to 1 W as stated in Eq. (15).

3 Radiation intensity

The radiation intensity is the time-averaged power flow per unit solid angle in a specified direction [9]. The electromagnetic fields and the time-averaged Poynting vector are required for the determination of the radiation intensity. First, the radiation in the space $0 < z < \infty$ is considered. Spherical coordinates (r, θ, ϕ) as defined by

$$x = r \sin \theta \cos \phi, \quad y = r \sin \theta \sin \phi, \quad \text{and} \quad z = r \cos \theta \tag{16}$$

are introduced. Since $\hat{z} = \hat{r} \cos \theta - \hat{\theta} \sin \theta$, the radial component of the time-averaged Poynting vector is found as $S_{r0}^{+}(r, \theta, \phi) = \cos \theta \, S_{z0}^{+}(x, y, z)$. The radiation intensity of the paraxial Gaussian beam is determined as

$$\Phi_0^{+}(\theta, \phi) = \lim_{kr \to \infty} r^2 S_{r0}^{+}(r, \theta, \phi) = \frac{\exp\left(-\frac{1}{2} k^2 w_0^2 \tan^2 \theta\right)}{2\pi f_0^2 \cos \theta} \tag{17}$$

where $f_0 = 1/k w_0$. $\Phi_0^{+}(\theta, \phi)$ is the radiation intensity of the fundamental Gaussian light beam specified by the vector potential as given by Eq. (11). Now, the radiation in the space $-\infty < z < 0$ is considered. Spherical coordinates (r, θ^-, ϕ) are defined with respect to the $-z$ axis. Then, in Eq. (16), the relation between z and r is changed as $-z = r\cos\theta^-$. Now, since $-\hat{z} = \hat{r}\cos\theta^- - \hat{\theta}^-\sin\theta^-$, the radial component of the time-averaged Poynting vector is found as $S_{r0}^-(r,\theta^-,\phi) = -\cos\theta^- S_{z0}^-(x,y,z)$. The radiation intensity of the paraxial Gaussian beam is obtained as

$$\Phi_0^-(\theta^-,\phi) = \lim_{kr\to\infty} r^2 S_{r0}^-(r,\theta^-,\phi) = \frac{\exp\left(-\frac{1}{2}k^2 w_0^2 \tan^2\theta^-\right)}{2\pi f_0^2 \cos\theta^-} \tag{18}$$

The radiation intensity distribution for $z < 0$ is found from the corresponding distribution for $z > 0$ by reflection about the $z = 0$ plane.

The radiation intensity given by Eq. (17) is cylindrically symmetrical about the $+z$ direction of propagation; that is, it is independent of ϕ. The radiation intensity has a maximum value of $1/2\pi f_0^2$ in the propagation direction $\theta = 0°$, which monotonically decreases as θ is increased and vanishes for $\theta = 90°$. The radiation intensity $\Phi_0^+(\theta,\phi)$ is plotted in Fig. 1 as a function of θ for $0° < \theta < 90°$ for three different values of kw_0, namely $kw_0 = 1.50$, 2.25, and 3.00. The peak of the radiation intensity pattern and the sharpness of the peak increase as kw_0 is increased.

The power carried by the paraxial Gaussian beam can also be deduced from the radiation intensity, as given by Eq. (17). The radiation intensity is multiplied by the element of solid angle $d\Omega = d\phi\sin\theta d\theta$ and integrated throughout the entire solid angle corresponding to $z > 0$ to obtain the total power as

$$P_0^+ = \int_0^{2\pi} d\phi \int_0^{\pi/2} d\theta \sin\theta\, \Phi_0^+(\theta,\phi) \tag{19}$$

The integration with respect to ϕ yields 2π. For the θ integration, the variable is changed to $\alpha = (0.5)^{1/2} kw_0\tan\theta$. Then, $d\alpha = (0.5)^{1/2} kw_0 d\theta(1 + 2f_0^2\alpha^2)$. The integral is evaluated as a power series in f_0^2 and the leading term alone is retained. Therefore, $d\alpha$ is approximated as

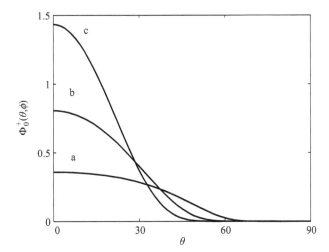

Fig. 1. Radiation intensity pattern of the fundamental electromagnetic Gaussian beam as a function of θ for $0° < \theta < 90°$. (a) $kw_0 = 1.50$, (b) $kw_0 = 2.25$, and (c) $kw_0 = 3.00$.

$da = (0.5)^{1/2} k w_0 d\theta$. When Eq. (17) is substituted into Eq. (19), the ϕ integration is carried out, and the variable θ is changed to the new variable α, it is obtained that

$$P_0^+ = \int_0^\infty \frac{da\,\alpha \exp(-\alpha^2)}{f_0^2 0.5(k w_0)^2} = 1 \tag{20}$$

The α integration is carried out to yield the result stated in Eq. (20). When the paraxial approximation of the result of the integrations in Eq. (19) is retained, Eq. (20) shows that the result obtained previously in Eq. (15) is reproduced.

Figure 1 shows that the width of the beam becomes smaller as $k w_0$ is increased. For $\theta = 0$, $\Phi_0^+(\theta, \phi) \to \infty$ for $k w_0 \to \infty$, and for $\theta \neq 0$, $\Phi_0^+(\theta, \phi) \to 0$ for $k w_0 \to \infty$. Also, since

$$P_0^+ = \int_0^{2\pi} d\phi \int_0^{\pi/2} d\theta \sin\theta\, \Phi_0^+(\theta, \phi) = 1 \tag{21}$$

the radiation intensity of the paraxial beam is singular as given by

$$\Phi_0^+(\theta, \phi) = \delta(1 - \cos\theta)/2\pi \quad \text{for } k w_0 \to \infty \tag{22}$$

For large and finite $k w_0$, $\Phi_0^+(\theta, \phi)$ spreads out in the θ direction. A quantitative measure of the beam width can be obtained. The square of the normalized width of the radiation intensity distribution of the paraxial Gaussian beam in the radial direction (that is, normal to the propagation direction) is given by $\sin^2\theta$. The average value of the square of the normalized width is determined as

$$\sigma_{x0}^2 = \langle \sin^2\theta \rangle = \frac{1}{P_0^+} \int_0^{2\pi} d\phi \int_0^{\pi/2} d\theta \sin\theta \sin^2\theta\, \Phi_0^+(\theta, \phi) = 2 f_0^2 \tag{23}$$

To obtain Eq. (23), in the same way as in Eq. (19), the integral is evaluated as a power series in f_0^2 and the leading term alone is retained for determining the results pertaining to the paraxial approximation. As $k w_0$ is increased, f_0^2 decreases, and the average of the normalized width in the radial direction decreases and the sharpness of the beam increases.

4 Radiative and reactive powers

$A_{x0}^\pm(x, y, z)$, as given by Eq. (11), is a continuous function of z at the secondary source plane $z = 0$; therefore, $A_{x0}^\pm(x, y, 0)$ is not the secondary source but only the input value of $A_{x0}^\pm(x, y, z)$. The electric field component $E_{x0}^\pm(x, y, z)$ and the magnetic field component $H_{y0}^\pm(x, y, z)$ are the only electromagnetic field components associated with the paraxial beam. Since $E_{x0}^\pm(x, y, z) = i k A_{x0}^\pm(x, y, z)$, $E_{x0}^\pm(x, y, z)$ is also a continuous function of z at $z = 0$. But $H_{y0}^\pm(x, y, z) = \pm i k A_{x0}^\pm(x, y, z)$; therefore, $H_{y0}^\pm(x, y, z)$ has a discontinuity at the secondary source plane $z = 0$. The discontinuity of the tangential component of the magnetic field is equivalent to an electric current sheet on the plane $z = 0$. The density of the electric current concentrated on the $z = 0$ plane is found from Eqs. (11) and (12) as

$$\mathbf{J}_0(x, y, z) = \hat{z} \times \hat{y}[H_{y0}^+(x, y, 0) - H_{y0}^-(x, y, 0)]\delta(z) = -\hat{x} 2N \exp[-(x^2 + y^2)/w_0^2]\delta(z) \tag{24}$$

The strength of the electric current source is given by the coefficient of $\delta(z)$ in Eq. (24).

The complex power is found from Eq. (D18) as

$$P_{C0} = -\frac{c}{2} \int_{-\infty}^{\infty} \int_{-\infty}^{\infty} \int_{-\infty}^{\infty} dx dy dz \, \mathbf{E}(x, y, z) \cdot \mathbf{J}^*(x, y, z)$$

$$= cN^2 \int_{-\infty}^{\infty} \int_{-\infty}^{\infty} dx dy \exp[-2(x^2 + y^2)/w_0^2] \tag{25}$$

The integrals in Eq. (25) are evaluated by the use of Eqs. (B1) and (B6), and N^2 is substituted from Eq. (2). The result is

$$P_{C0} = cN^2 \pi w_0^2 / 2 = 2 \text{ W} \tag{26}$$

The reactive power is zero. The real power is equal to 2 W, of which 1 W flows in the $+z$ direction and the other 1 W flows in the $-z$ direction.

The complex power is evaluated for a few other types of paraxial beams. For every paraxial beam, the reactive power vanishes. Thus, it appears that the vanishing of the reactive power for the paraxial beams is a general result. Therefore, there is a need for the full-wave generalization of paraxial beams since a full characterization of laser output with respect to the complex power is desirable.

5 Beam spreading on propagation

The propagation characteristics are the same in the $\pm z$ directions, as is seen from Eq. (14); therefore, $\pm$ signs are omitted. The beam spreads out in the transverse (x, y) directions as it propagates in the longitudinal $(\pm z)$ direction. From Eqs. (14) and (2), the magnitude of the Poynting vector is normalized to yield

$$\pm S_{z0}^{\pm}(x, y, z) \frac{\pi w_0^2}{2} = S_{n0}(\rho, z) \tag{27}$$

where

$$S_{n0}(\rho, z) = \frac{1}{(1 + z^2/b^2)} \exp\left[-\frac{2\rho^2}{w_0^2(1 + z^2/b^2)}\right] \tag{28}$$

and

$$x = \rho \cos \phi \quad \text{and} \quad y = \rho \sin \phi \tag{29}$$

The beam is cylindrically symmetrical about the propagation direction. Along the beam axis, $\rho = 0$, $S_{n0}(\rho, z)$ decreases as $|z|$ increases from $|z| = 0$. For ρ sufficiently greater than w_0, as $|z|/b$ increases, the factor outside the exponent decreases but the exponential term increases, causing $S_{n0}(\rho, z)$ to increase. Consequently, the beam becomes less sharp, spreads out in the transverse direction (ρ), eventually gets sufficiently diffused, and loses its beam-like character. In Fig. 2, $S_{n0}(\rho, z)$ is plotted for ρ in the range $0 < \rho/w_0 < 2.5$, for $\phi = 0$ and $\phi = \pi$,

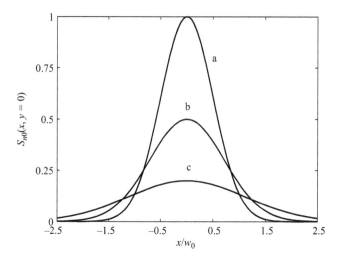

Fig. 2. Normalized Poynting vector $S_{n0}(x, y = 0)$ as a function of x/w_0 in the range $-2.5 < x/w_0 < 2.5$ and for (a) $z/b = 0$, (b) $z/b = 1$, and (c) $z/b = 2$.

and for three values of $z/b = 0$, 1, and 2. For $\phi = 0$, $y = 0$, x is positive and is in the range $0 < x/w_0 < 2.5$, and for $\phi = \pi$, $y = 0$, x is negative and is in the range $-2.5 < x/w_0 < 0$. Figure 2 shows that, on propagation, the beam spreads out and loses its beam-like property.

If w_0 is decreased, Eq. (1) shows that the input distribution becomes sharper. If w_0 is decreased, b becomes smaller at a greater rate and the same amount of spreading occurs for a smaller value of z. Consequently, sharper input distribution leads to a more rapid spreading of the beam. The particular solution of the approximate paraxial wave equation depends on the chosen input field distribution. Different input distributions lead to different types of paraxial beams. The spreading characteristics of different paraxial beams are different. One area of research on electromagnetic paraxial beams is to find a particular input field distribution that leads to no or very small spreading on propagation as a consequence of the Fresnel diffraction.

6 Magnetic current density

The discontinuity of the tangential component of the magnetic field at the secondary source plane is equivalent to an electric current sheet on the plane $z = 0$. By duality, the discontinuity of the tangential component of the electric field is equivalent to a magnetic current sheet on the plane $z = 0$. The fundamental Gaussian light beam treated in the preceding sections has its electric field polarized in the x direction. For generating the orthogonal polarization, namely the fundamental Gaussian light beam with its magnetic field polarized in the x direction, the paraxial approximation of the x component of the electric vector potential on the input plane $z = 0$ as given by

$$F_{x0}^{\pm}(x, y, 0) = \frac{N}{ik} \exp\left(-\frac{x^2 + y^2}{w_0^2}\right) \tag{30}$$

is required. As before, $F_0^\pm(x,y,z)$ can be determined as follows:

$$F_{x0}^\pm(x,y,z) = \exp(\pm ikz)\frac{N}{ik}q_\pm^2 \exp\left[-\frac{q_\pm^2(x^2+y^2)}{w_0^2}\right] \qquad (31)$$

The resulting electromagnetic fields are evaluated by the use of Eqs. (D26)–(D29). The magnetic field component $H_{x0}^\pm(x,y,z)$ and the electric field component $E_{y0}^\pm(x,y,z)$ are the only electromagnetic field components associated with the paraxial beam. From Eqs. (D28) and (D27), respectively, it is found that $H_{x0}^\pm(x,y,z) = ikF_{x0}^\pm(x,y,z)$ and $E_{y0}^\pm(x,y,z) = \pm(-1)ikF_{x0}^\pm(x,y,z)$. Therefore, $F_{x0}^\pm(x,y,z)$ and $H_{x0}^\pm(x,y,z)$ are continuous across the secondary source plane $z = 0$. But $E_{y0}^\pm(x,y,z)$ has a discontinuity at the secondary source plane. The discontinuity of $E_{y0}^\pm(x,y,z)$ is equivalent to a magnetic current sheet on the plane $z = 0$. A comparison of Eqs. (D8) and (D9) shows that in the manner in which the magnetic current density is introduced in Maxwell's equations, there is a minus sign associated with the magnetic current density. Hence, the density of the magnetic current concentrated on the $z = 0$ plane is given by

$$\mathbf{J}_0(x,y,z) = -\hat{z} \times \hat{y}[E_{y0}^+(x,y,0) - E_{y0}^-(x,y,0)]\delta(z) \qquad (32)$$

Equations (D18) and (D22) show that in the expression of the complex power associated with the electric current density and the magnetic current density, the electric field and the electric current density are replaced, respectively, without any change in sign, with the magnetic field and the magnetic current density.

7 Some applications and limitations

The fundamental Gaussian beam is used in the modeling of the output of a laser in free space propagation. The width of the beam in the radial direction varies on propagation. The minimum value of the width of the beam is the waist of the beam. The waist is a characteristic parameter of the beam. There are many applications for the approximate fundamental Gaussian beam. In the study of reflection, refraction, diffraction, and scattering, normally only a plane electromagnetic wave is used for the incident wave. Because infinite energy is required for its launching, a plane electromagnetic wave is not physically realizable. But the corresponding paraxial electromagnetic beam requires only a finite energy for its launching and is a good approximation to the plane wave to use for the incident wave. Using a paraxial beam for the incident wave, reflection and refraction at an interface, diffraction and scattering by planar and parabolic reflectors, by a half-plane, and by a cylinder have been investigated [10–14]. The use of a paraxial beam instead of a plane wave has revealed new features in the classical phenomena of reflection, refraction, diffraction, and scattering of electromagnetic waves.

For propagation in an optical system of lenses with an axis of symmetry in the propagation direction, the light wave is modeled by the fundamental Gaussian beam [8]. The Gaussian beam can be transformed by a lens. For example, the location and the size of the beam waist can be made different from those of the incident Gaussian beam. Consider a coaxial and periodic system of lenses in the propagation direction. The fundamental

Gaussian beam from the laser is launched on the first lens. The beam, which spreads out as it reaches the first lens, is focused. Then the beam converges until it reaches the waist size and then spreads out again. The system is arranged such that the beam waist is located halfway between the two lenses. The beam reaches the second lens in a manner exactly the same as for the first lens. The process between the first two lenses is repeated between the second pair of lenses, and then between successive pairs of lenses. Thus, the coaxial and periodic system of lenses guides the fundamental Gaussian beam in the propagation direction. The propagation of the fundamental Gaussian beam through lens-like media with loss or gain variation has also been investigated [15–17].

Fabry–Perot interferometers, used for the spectrum analysis of light, consist of spherical mirror cavities formed by two concave mirrors facing each other. Light inside the cavity is modeled by the fundamental Gaussian beam multiply reflected by the two end mirrors. Sharp cavity resonances occur at multiple frequencies. The sharp frequency response of the light transmitted from the cavity is used for the measurement of the frequency spectrum of the light [8].

An entire area in modern optics, namely Fourier optics, was developed by the use of the fundamental scalar Gaussian beam for the description of the laser light [3, 8]. Another area in contemporary optics, coherent wave optics, has emerged by the use of the fundamental scalar Gaussian beam for the description of the light wave [7, 18, 19]. This area includes a treatment of the propagation properties of light wave whose amplitude fluctuates randomly on a time scale that is large in comparison with the wave period.

The radiative power of lasers is propagated far from the source, and the reactive power remains close to the source [20, 21]. Therefore, the radiative and the reactive powers are related to far-field and near-field optics, respectively. The growing interest in near-field optics has created the need for the determination of the reactive power of lasers. The linearly polarized paraxial electromagnetic Gaussian beams have no reactive power. Consequently, a realistic characterization of light sources with regard to the reactive power requires a full-wave generalization of the various paraxial electromagnetic Gaussian beams. This monograph is devoted to the full-wave generalization of the various commonly occurring paraxial electromagnetic Gaussian beams.

References

1. H. Kogelnik and T. Li, "Laser beams and resonators," *Appl. Opt.* **5**, 1550–1567 (1966).

2. A. Yariv, *Quantum Electronics*, 2nd ed. (Wiley, New York, 1967), Chap. 6.

3. J. W. Goodman, *Introduction to Fourier Optics* (McGraw-Hill, New York, 1968), Chaps. 3 and 4.

4. D. Marcuse, *Light Transmission Optics* (Van Nostrand Reinhold, New York, 1972), Chap. 6.

5. H. A. Haus, *Waves and Fields in Optoelectronics* (Prentice-Hall, Englewood Cliffs, NJ, 1984), Chaps. 4, 5, and 11.

6. A. E. Siegman, *Lasers* (University Science, Mill Valley, CA, 1986), Sects. 16.1 and 17.1.

7. L. Mandel and E. Wolf, *Optical Coherence and Quantum Optics* (Cambridge University Press, New York, 1995), pp. 263–287.

8. K. Iizuka, *Elements of Photonics* (Wiley-Interscience, New York, 2002).

9. S. R. Seshadri, *Fundamentals of Transmission Lines and Electromagnetic Fields* (Addison-Wesley, Reading, MA, 1971), pp. 468–470.

10. E. Gowan and G. A. Deschamps, "Quasi-optical approaches to the diffraction and scattering of Gaussian beams," University of Illinois, Antenna Laboratory, Report 70-5, Urbana-Champaign, Illinois, 1970.

11. J. W. Ra, H. Bertoni, and L. B. Felsen, "Reflection and transmission of beams at dielectric interfaces," *SIAM J. Appl. Math.* **24**, 396–412 (1973).

12. W.-Y. D. Wang and G. A. Deschamps, "Application of complex ray tracing to scattering problems," *Proc. IEEE* **62**, 1541–1551 (1974).

13. A. C. Green, H. L. Bertoni, and L. B. Felsen, "Properties of the shadow cast by a half-screen when illuminated by a Gaussian beam," *J. Opt. Soc. Am.* **69**, 1503–1508 (1979).

14. G. A. Suedan and E. V. Jull, "Beam diffraction by planar and parabolic reflectors," *IEEE Trans. Antennas Propag.* **39**, 521–527 (1991).

15. G. Goubau and J. Schwering, "On the guided propagation of electromagnetic beam waves," *IRE Trans. Antennas and Propag.* **AP-9**, 248–256 (1961).

16. H. Kogelnik, "On the propagation of Gaussian beams of light through lens like media including those with loss or gain variation," *Appl. Opt.* **4**, 1562–1569 (1965).

17. P. K. Tien, J. P. Gordon, and J. R. Whinnery, "Focusing of a light beam of Gaussian field distribution in continuous and periodic lens like media," *Proc. IEEE*, **53**, 129–136 (1965).

18. E. Wolf, *Introduction to the Theory of Coherence and Polarization of Light* (Cambridge University Press, Cambridge, UK, 2007).

19. J. W. Goodman, "*Statistical Optics* (Wiley, New York, 1985).

20. S. R. Seshadri, "Constituents of power of an electric dipole of finite size," *J. Opt. Soc. Am. A* **25**, 805–810 (2008).

21. S. R. Seshadri, "Power of a simple electric multipole of finite size," *J. Opt. Soc. Am. A* **25**, 1420–1425 (2008).

CHAPTER 2

Fundamental Gaussian wave

The secondary source for the approximate paraxial beams and the exact full waves is a current sheet that is situated on the plane $z = 0$. The beams and the waves generated by the secondary source propagate out in the $+z$ direction in the space $0 < z < \infty$ and in the $-z$ direction in the space $-\infty < z < 0$. The response of the electric current source given by Eq. (1.24) obtained in the paraxial approximation is the fundamental Gaussian beam. The same current source for the full Helmholtz wave equation yields the fundamental Gaussian wave. For the electric current source given by Eq. (1.24), the Helmholtz wave equation is solved to obtain the exact vector potential. The electromagnetic fields are derived, the radiation intensity distribution is determined, and its characteristics are analyzed. The time-averaged power transported by the fundamental Gaussian wave in the $\pm z$ direction is obtained. The complex power is evaluated and from there the reactive power is found. The time-averaged power carried by the fundamental Gaussian wave in the $\pm z$ direction increases, reaches a maximum greater than 1, decreases, and approaches the value of 1 corresponding to the fundamental Gaussian beam as the parameter kw_0 is increased. The reactive power does not vanish for the fundamental Gaussian wave. The reactive power decreases, reaches zero, decreases further, reaches a minimum, then increases and approaches zero (that is, the limiting value for the corresponding fundamental Gaussian beam) as kw_0 is increased.

1 Exact vector potential

The electric current density given by Eq. (1.24) is in the x direction; therefore, the exact vector potential $A_x^{\pm}(x, y, z)$ is also in the x direction. From Eqs. (1.24) and (D25), $A_x^{\pm}(x, y, z)$ is found to be governed by the following inhomogeneous Helmholtz wave equation:

$$\left(\frac{\partial^2}{\partial x^2} + \frac{\partial^2}{\partial y^2} + \frac{\partial^2}{\partial z^2} + k^2 \right) A_x^{\pm}(x, y, z) = 2N \exp\left(-\frac{x^2 + y^2}{w_0^2} \right) \delta(z) \tag{1}$$

$A_x^{\pm}(x, y, z)$ is expressed in terms of $\overline{A}_x^{\pm}(p_x, p_y, z)$, the two-dimensional Fourier transform of $A_x^{\pm}(x, y, z)$, by the use of Eq. (A17); the exponential function on the right-hand side of Eq. (1)

is stated as the inverse Fourier transform of a function by the use of Eqs. (1.1) and (1.3). Then the following one-dimensional governing equation for $\overline{A}_x^{\pm}(p_x, p_y, z)$ is obtained:

$$\left(\frac{\partial^2}{\partial z^2} + \zeta^2\right)\overline{A}_x^{\pm}(p_x, p_y, z) = 2N\pi w_0^2 \exp\left[-\pi^2 w_0^2\left(p_x^2 + p_y^2\right)\right]\delta(z) \tag{2}$$

where ζ is positive real or positive imaginary as given by

$$\begin{aligned}\zeta &= \left[k^2 - 4\pi^2\left(p_x^2 + p_y^2\right)\right]^{1/2} \quad \text{for } k^2 > 4\pi^2\left(p_x^2 + p_y^2\right)\\ &= i\left[4\pi^2\left(p_x^2 + p_y^2\right) - k^2\right]^{1/2} \quad \text{for } k^2 < 4\pi^2\left(p_x^2 + p_y^2\right)\end{aligned} \tag{3}$$

By the use of Eqs. (A19)–(A26), Eq. (2) is solved to yield

$$\overline{A}_x^{\pm}(p_x, p_y, z) = \frac{N}{i}\pi w_0^2 \exp\left[-\pi^2 w_0^2\left(p_x^2 + p_y^2\right)\right]\zeta^{-1}\exp(i\zeta|z|) \tag{4}$$

The inverse Fourier transform of Eq. (4) yields the exact vector potential as

$$\begin{aligned}A_x^{\pm}(x, y, z) = \frac{N}{i}\pi w_0^2 \int_{-\infty}^{\infty}\int_{-\infty}^{\infty} dp_x dp_y \exp\left[-i2\pi\left(p_x x + p_y y\right)\right]\\ \times \exp\left[-\pi^2 w_0^2\left(p_x^2 + p_y^2\right)\right]\zeta^{-1}\exp(i\zeta|z|)\end{aligned} \tag{5}$$

The paraxial approximation corresponds to $4\pi^2(p_x^2 + p_y^2)/k^2 \ll 1$ [see Eq. (C17)]. When ζ, given by Eq. (3), is expanded into a power series in $4\pi^2(p_x^2 + p_y^2)/k^2$, the first two terms are given by

$$\zeta = k - \pi^2 w_0^2\left(p_x^2 + p_y^2\right)/b \tag{6}$$

In Eq. (5), if ζ in the amplitude is replaced by the first term in Eq. (6) and ζ in the phase is replaced by the first two terms in Eq. (6), the paraxial approximation to Eq. (5) is found as

$$\begin{aligned}A_{x0}^{\pm}(x, y, z) = \frac{N}{ik}\exp(\pm ikz)\pi w_0^2 \int_{-\infty}^{\infty}\int_{-\infty}^{\infty} dp_x dp_y \exp\left[-i2\pi\left(p_x x + p_y y\right)\right]\\ \times \exp\left[-\frac{\pi^2 w_0^2\left(p_x^2 + p_y^2\right)}{q_{\pm}^2}\right]\end{aligned} \tag{7}$$

where $q_{\pm}$ is defined by Eq. (1.9). The integrals in Eq. (7) are evaluated by the use of Eqs. (B1) and (B6) with the result that

$$A_{x0}^{\pm}(x, y, z) = \exp(\pm ikz)\frac{N}{ik}q_{\pm}^2 \exp\left[-\frac{q_{\pm}^2(x^2 + y^2)}{w_0^2}\right] \tag{8}$$

The exact vector potential as given by Eq. (5) reduces correctly to the paraxial beam result in the appropriate limit.

2 Exact electromagnetic fields

The electromagnetic fields are derived by substituting $A_x^{\pm}(x, y, z)$ from Eq. (5) into Eqs. (D30)–(D33). Thus, $H_x^{\pm}(x, y, z) \equiv 0$ and all the other field components are excited. For finding the total power radiated and the radiation intensity distributions in $z > 0$ and $z < 0$, only the $\pm z$ component of the time-averaged Poynting vector is needed. For this purpose, since $H_x^{\pm}(x, y, z) \equiv 0$, only the field components $E_x^{\pm}(x, y, z)$ and $H_y^{\pm}(x, y, z)$ are required. For evaluating the complex power, since the electric current density is in the x direction, only the x component of the electric field, namely $E_x^{\pm}(x, y, z)$, is needed. Therefore, for our present purposes, only the field components $E_x^{\pm}(x, y, z)$ and $H_y^{\pm}(x, y, z)$ are required. Interestingly enough, these are the only field components that exist in the paraxial approximation. The electromagnetic field components $E_x^{\pm}(x, y, z)$ and $H_y^{\pm}(x, y, z)$ are found from Eqs. (D30) and (D33) as

$$E_x^{\pm}(x, y, z) = ik\left(1 + \frac{1}{k^2}\frac{\partial^2}{\partial x^2}\right)A_x^{\pm}(x, y, z) \tag{9}$$

and

$$H_y^{\pm}(x, y, z) = \frac{\partial A_x^{\pm}(x, y, z)}{\partial z} \tag{10}$$

Substituting $A_x^{\pm}(x, y, z)$ from Eq. (5) into Eqs. (9) and (10) leads to

$$E_x^{\pm}(x, y, z) = N\pi k w_0^2 \int_{-\infty}^{\infty} \int_{-\infty}^{\infty} dp_x dp_y \exp[-i2\pi(p_x x + p_y y)]$$
$$\times \left(1 - \frac{4\pi^2 p_x^2}{k^2}\right)\exp\left[-\pi^2 w_0^2\left(p_x^2 + p_y^2\right)\right]\zeta^{-1}\exp(i\zeta|z|) \tag{11}$$

and

$$H_y^{\pm}(x, y, z) = \pm N\pi w_0^2 \int_{-\infty}^{\infty} \int_{-\infty}^{\infty} d\bar{p}_x d\bar{p}_y \exp\left[-i2\pi\left(\bar{p}_x x + \bar{p}_y y\right)\right]$$
$$\times \exp\left[-\pi^2 w_0^2\left(\bar{p}_x^2 + \bar{p}_y^2\right)\right]\exp(i\bar{\zeta}|z|) \tag{12}$$

where $\bar{\zeta}$ is the same as ζ with p_x and p_y changed to $\bar{p}_x$ and $\bar{p}_y$, respectively.

The time-averaged power flow per unit area in the $\pm z$ direction is determined by the use of Eq. (1.13). The use of Eqs. (11) and (12) in Eq. (1.13) yields

$$\pm S_z^{\pm}(x, y, z) = \pm\frac{c}{2}\text{Re}\left\{N\pi k w_0^2 \int_{-\infty}^{\infty}\int_{-\infty}^{\infty} dp_x dp_y\right.$$
$$\times \exp[-i2\pi(p_x x + p_y y)]\left(1 - \frac{4\pi^2 p_x^2}{k^2}\right)$$
$$\times \exp\left[-\pi^2 w_0^2\left(p_x^2 + p_y^2\right)\right]\zeta^{-1}\exp(i\zeta|z|)$$
$$\times (\pm)N\pi w_0^2 \int_{-\infty}^{\infty}\int_{-\infty}^{\infty} d\bar{p}_x d\bar{p}_y \exp[i2\pi(\bar{p}_x x + \bar{p}_y y)]$$
$$\left.\times \exp\left[-\pi^2 w_0^2\left(\bar{p}_x^2 + \bar{p}_y^2\right)\right]\exp(-i\bar{\zeta}^*|z|)\right\} \tag{13}$$

The time-averaged power $P^\pm$ transported by the fundamental Gaussian wave in the $\pm z$ direction is found by integrating $[\pm S_z^\pm(x, y, z)]$ with respect to x and y across the entire transverse plane, with the result that

$$P^\pm = \int_{-\infty}^{\infty} \int_{-\infty}^{\infty} dx dy [\pm S_z^\pm(x, y, z)] \tag{14}$$

From Eq. (13), $\pm S_z^\pm(x, y, z)$ is substituted into Eq. (14) and the integrations with respect to x and y are performed first, yielding $\delta(p_x - \bar{p}_x)$ and $\delta(p_y - \bar{p}_y)$. The integrations with respect to $\bar{p}_x$ and $\bar{p}_y$ are carried out next. The result is

$$P^\pm = \frac{c}{2} N^2 \pi^2 k w_0^4 \mathrm{Re} \int_{-\infty}^{\infty} \int_{-\infty}^{\infty} dp_x dp_y \left(1 - \frac{4\pi^2 p_x^2}{k^2}\right)$$
$$\times \exp\left[-2\pi^2 w_0^2 \left(p_x^2 + p_y^2\right)\right] \zeta^{-1} \exp[\pm iz(\zeta - \zeta^*)] \tag{15}$$

From Eq. (1.2), the value of N^2 is substituted. Then, the value of $P^\pm$ in the limit of the paraxial approximation equals $P_0^\pm = 1$ W. In other words, the normalization is such that the time-averaged power transported by the fundamental Gaussian beam in the $\pm z$ direction is equal to 1 W. The integration variables p_x and p_y are changed as follows:

$$2\pi p_x = p \cos \phi \quad 2\pi p_y = p \sin \phi \tag{16}$$

Then, $P^\pm$ given by Eq. (15) is transformed as

$$P^\pm = \frac{w_0^2}{2\pi} \mathrm{Re} \int_0^{\infty} dp p \int_0^{2\pi} d\phi \left(1 - \frac{p^2 \cos^2 \phi}{k^2}\right) \exp\left(-\frac{w_0^2 p^2}{2}\right) \frac{1}{\xi} \exp[\pm ikz(\xi - \xi^*)] \tag{17}$$

where, in accordance with Eq. (3), $\xi = \zeta/k$ is either positive real or positive imaginary, as given by

$$\xi = (1 - p^2/k^2)^{1/2} \quad \text{for } 0 < p < k$$
$$= i(p^2/k^2 - 1)^{1/2} \quad \text{for } k < p < \infty \tag{18}$$

For $k < p < \infty$, ξ is imaginary; then, the integrand is imaginary and has no real part. Hence, the contribution to $P^\pm$ for p in the range $k < p < \infty$ vanishes. Another change of variable is now introduced as

$$p = k \sin \theta^+ \tag{19}$$

Then it is found from Eqs. (18) and (19) that

$$\xi = |\cos \theta^+| \tag{20}$$

For $p = 0$, $\theta^+ = 0$ or π, and for $p = k$, $\theta^+ = \pi/2$. Therefore, θ^+ has two possible ranges of values. For $z > 0$, $0 < \theta^+ < \pi/2$ and $\xi = \cos \theta^+$. Then P^+, given by Eq. (17), is simplified and expressed as

$$P^+ = \int_0^{\pi/2} d\theta^+ \sin \theta^+ \int_0^{2\pi} d\phi \Phi(\theta^+, \phi) \tag{21}$$

where

$$\Phi(\theta^+,\phi) = \frac{(1 - \sin^2\theta^+ \cos^2\phi)}{2\pi f_0^2} \exp\left(-\frac{1}{2}k^2 w_0^2 \sin^2\theta^+\right) \tag{22}$$

and $k^2 w_0^2 = f_0^{-2}$. For $z < 0$, $\pi > \theta^+ > \pi/2$ and $\xi = -\cos\theta^+$. Then P^-, given by Eq. (17), is simplified as

$$P^- = -\int_\pi^{\pi/2} d\theta^+ \sin\theta^+ \int_0^{2\pi} d\phi\,\Phi(\theta^+,\phi) \tag{23}$$

Let

$$\theta^- = \pi - \theta^+ \tag{24}$$

Then it is found from Eq. (22) that

$$\Phi(\pi - \theta^-,\phi) = \Phi(\theta^-,\phi) \tag{25}$$

In view of Eqs. (24) and (25), P^- given by Eq. (23) is transformed as

$$P^- = \int_0^{\pi/2} d\theta^- \sin\theta^- \int_0^{2\pi} d\phi\,\Phi(\theta^-,\phi) \tag{26}$$

For $z > 0$, if θ^+ is defined with respect to $+z$ axis, and for $z < 0$, if θ^- is defined with respect to $-z$ axis, the expressions for P^+ as given by Eqs. (21) and (22) for $z > 0$ and for P^- as given by Eqs. (25) and (26) for $z < 0$ are the same. Therefore, $P^+ = P^-$.

The fundamental Gaussian wave is seen from Eq. (22) to be characterized by one wave parameter kw_0. Equation (22) is substituted into Eq. (21), the ϕ integration is performed analytically, and the θ^+ integration is carried out numerically to obtain $P^\pm$ as a function of kw_0. In Fig. 1, $P^\pm$ is shown as a function of kw_0 for $1 < kw_0 < 5$. $P^\pm$ increases, reaches a maximum value greater than 1, decreases, and approaches the limiting value of $P_0^\pm = 1$ W corresponding to the fundamental Gaussian beam as kw_0 is increased.

The ratio of the time-averaged power $P_0^\pm$ in the fundamental Gaussian beam to the time-averaged power $P^\pm$ in the fundamental Gaussian wave is valuable for determining the quality of the paraxial beam approximation to the exact full wave for different values of the wave parameter kw_0 [1]. Since, for a large range of variation of kw_0, $P^\pm/P_0^\pm$ approaches 1 as kw_0 is increased, the quality of the paraxial beam approximation to the full wave improves, in general, as kw_0 is increased.

3 Radiation intensity

For $z > 0$, it follows from Eqs. (21) and (22) that $\Phi(\theta^+,\phi)$ is the radiation intensity distribution of the fundamental Gaussian wave. Equations (25) and (26) show that $\Phi(\theta^-,\phi)$ is the radiation intensity distribution for $z < 0$. Here, θ^- for $z < 0$ is the angle defined with respect to the $-z$ axis in the same way that θ^+ for $z > 0$ is the corresponding angle defined

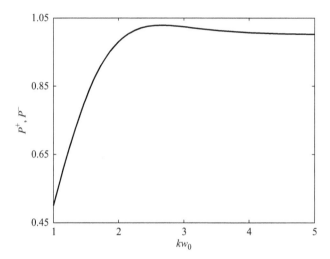

Fig. 1. Time-averaged power $P^\pm$ in watts transported by the fundamental Gaussian wave in the $\pm z$ directions as a function of kw_0 for $1 < kw_0 < 5$. In the limit of the paraxial approximation, $P^\pm$ equals $P_0^\pm = 1$ W.

with respect to the $+z$ axis. As a consequence, the radiation intensity distribution for $z < 0$ is obtained by reflection about the $z = 0$ plane of the corresponding radiation intensity distribution for $z > 0$. Hence, only the characteristics of the radiation intensity distribution for $z > 0$ are discussed.

The radiation intensity $\Phi(\theta^+, \phi)$ is not cylindrically symmetrical; that is, it is a function of ϕ. But $\Phi(\theta^+, \phi)$ has reflection symmetries about the $\phi = (0°, 180°)$ and $\phi = (90°, 270°)$ planes; therefore, it is sufficient to examine the variation of $\Phi(\theta^+, \phi)$ for one quadrant, namely for $0° < \phi < 90°$. In every azimuthal plane $\phi =$ constant, $\Phi(\theta^+, \phi)$ has the value $1/2\pi f_0^2$ for $\theta^+ = 0°$, decreases monotonically as θ^+ increases, and reaches a minimum value for $\theta^+ = 90°$; this minimum value increases from 0 for $\phi = 0°$ to a maximum value for $\phi = 90°$. The beam width of the fundamental Gaussian wave has a minimum value for $\phi = 0°$, increases continuously as ϕ is increased, and reaches a maximum value for $\phi = 90°$.

In Fig. 2, the radiation intensity pattern $\Phi(\theta^+, \phi)$ of the fundamental Gaussian wave is shown as functions of θ^+ for $0° < \theta^+ < 90°$, $kw_0 = 1.563$, and for four different values of ϕ, namely $\phi = 0°, 30°, 60°$, and $90°$. The cylindrically symmetric radiation intensity pattern $\Phi_0(\theta^+, \phi)$ of the corresponding fundamental Gaussian beam is also included for comparison purposes. The normalization is such that the time-averaged power $P_0^\pm$ in the fundamental Gaussian beam propagating in the $+z$ or the $-z$ direction is 1 W. Figure 2 reveals that $\Phi_0(\theta^+, \phi)$ is a reasonably good approximation to $\Phi(\theta^+, \phi)$ for all ϕ and for $0° < \theta^+ < 90°$. $\Phi(\theta^+, \phi)$ is not cylindrically symmetrical, but the paraxial approximation $\Phi_0(\theta^+, \phi)$ is cylindrically symmetrical. The quality of the agreement between $\Phi(\theta^+, \phi)$ and $\Phi_0(\theta^+, \phi)$ decreases as ϕ is decreased from $90°$ to $0°$. Thus, the overall differences between $\Phi(\theta^+, \phi)$ and $\Phi_0(\theta^+, \phi)$ are largest for $\phi = 0°$.

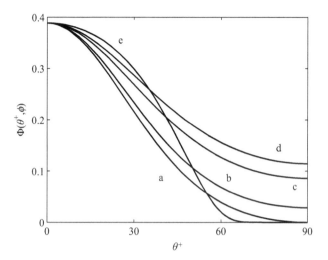

Fig. 2. Radiation intensity pattern $\Phi(\theta^+, \phi)$ of the fundamental Gaussian wave for curves (a) $\phi = 0°$, (b) $\phi = 30°$, (c) $\phi = 60°$, (d) $\phi = 90°$, and (e) cylindrically symmetric radiation intensity pattern, $\Phi_0(\theta^+, \phi)$ of the corresponding fundamental Gaussian beam as functions of θ^+ for $0° < \theta^+ < 90°$. Other parameters are $kw_0 = 1.563$; the total power in the fundamental Gaussian beam is 2 W, and that in the fundamental Gaussian wave is 1.686 W.

4 Radiative and reactive powers

The integral representation of $E_x^{\pm}(x, y, z)$ is obtained in Eq. (11). Similarly, the integral representation of the source electric current density given by Eq. (1.24) is found as

$$\mathbf{J}_0(x, y, z) = -\hat{x} 2N\pi w_0^2 \delta(z) \int_{-\infty}^{\infty} \int_{-\infty}^{\infty} d\bar{p}_x d\bar{p}_y$$
$$\times \exp\left[-i2\pi\left(\bar{p}_x x + \bar{p}_y y\right)\right] \exp\left[-\pi^2 w_0^2\left(\bar{p}_x^2 + \bar{p}_y^2\right)\right] \tag{27}$$

The complex power is determined from Eq. (D18) as

$$P_C = -\frac{c}{2} \int_{-\infty}^{\infty} \int_{-\infty}^{\infty} \int_{-\infty}^{\infty} dx\,dy\,dz\, \mathbf{E}(x, y, z) \cdot \mathbf{J}_0^*(x, y, z) \tag{28}$$

When Eqs. (11) and (27) are substituted into Eq. (28) and the integration with respect to z is performed, the result is

$$P_C = \frac{c}{2} \int_{-\infty}^{\infty} \int_{-\infty}^{\infty} dx\,dy\, N\pi k w_0^2 \int_{-\infty}^{\infty} \int_{-\infty}^{\infty} dp_x dp_y \left(1 - \frac{4\pi^2 p_x^2}{k^2}\right)$$
$$\times \exp[-i2\pi(p_x x + p_y y)] \exp\left[-\pi^2 w_0^2\left(p_x^2 + p_y^2\right)\right] \zeta^{-1}$$
$$\times 2N\pi w_0^2 \int_{-\infty}^{\infty} \int_{-\infty}^{\infty} d\bar{p}_x d\bar{p}_y \exp[-i2\pi(\bar{p}_x x + \bar{p}_y y)]$$
$$\times \exp\left[-\pi^2 w_0^2\left(\bar{p}_x^2 + \bar{p}_y^2\right)\right] \tag{29}$$

The same procedure that was used for simplifying Eq. (14) is used. Then, P_C given by Eq. (29) reduces as

$$P_C = cN^2\pi^2 kw_0^4 \int_{-\infty}^{\infty}\int_{-\infty}^{\infty} dp_x dp_y \left(1 - \frac{4\pi^2 p_x^2}{k^2}\right) \exp[-2\pi^2 w_0^2 (p_x^2 + p_y^2)]\zeta^{-1} \tag{30}$$

Equation (30) is simplified in the same manner as Eq. (14) to yield that

$$P_C = P_{re} + iP_{im} = \frac{w_0^2}{\pi} \int_0^{\infty} dpp \int_0^{2\pi} d\phi \left(1 - \frac{p^2 \cos^2\phi}{k^2}\right) \exp\left(-\frac{w_0^2 p^2}{2}\right)\frac{1}{\xi} \tag{31}$$

The value of the integral with respect to p is real only for $0 < p < k$. Then the use of Eqs. (19)–(22) shows that

$$P_{re} = 2P^+ = P^+ + P^- \tag{32}$$

Therefore, the real power is equal to the total time-averaged power transported by the fundamental Gaussian wave in the $+z$ and the $-z$ directions. The real power (that is, the time-averaged power generated by the current source) can be determined by volume integration over the entire source current distribution. The time-averaged value of the power created by the current sources can also be obtained from the time-averaged power flowing out of a large surface enclosing the current sources. As proved in Eq. (D13) in general, and as verified in Eq. (32) for the special case of the fundamental Gaussian wave, the two procedures yield identical results.

The value of the integral with respect to p is imaginary for $k < p < \infty$. Then, after the integration with respect to ϕ is carried out, the use of Eq. (18) in Eq. (31) enables us to find the reactive power P_{im} as

$$P_{im} = -2w_0^2 \int_k^{\infty} dpp \left(1 - \frac{p^2}{2k^2}\right) \exp\left(-\frac{w_0^2 p^2}{2}\right)\left(\frac{p^2}{k^2} - 1\right)^{-1/2} \tag{33}$$

The variable of integration is changed as

$$p^2 = k^2(1 + \tau^2) \tag{34}$$

Then, P_{im} given by Eq. (33) simplifies as

$$P_{im} = -k^2 w_0^2 \exp\left(-\frac{1}{2}k^2 w_0^2\right)\int_0^{\infty} d\tau (1 - \tau^2) \exp\left(-\frac{1}{2}k^2 w_0^2\tau^2\right) \tag{35}$$

The integral in Eq. (35) is evaluated and the reactive power is obtained as

$$P_{im} = -\left(\frac{\pi}{2}\right)^{1/2} kw_0 \left(1 - \frac{1}{k^2 w_0^2}\right)\exp\left(-\frac{1}{2}k^2 w_0^2\right) \tag{36}$$

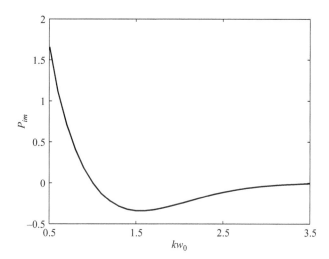

Fig. 3. Reactive power P_{im} of the fundamental Gaussian wave as a function of kw_0 for $0.5 < kw_0 < 3.5$. The total power in the fundamental Gaussian beam is 2 W.

For the limiting case ($kw_0 \to \infty$) of the fundamental Gaussian beam, the reactive power is zero. For the fundamental Gaussian wave, the reactive power does not vanish. The reactive power can be positive or negative. For $kw_0 < 1$, $P_{im} > 0$, and for $kw_0 > 1$, $P_{im} < 0$. There is a resonance in the sense that $P_{im} = 0$ for $kw_0 = 1$. In Fig. 3, P_{im} is shown as a function of the wave parameter kw_0 for $0.5 < kw_0 < 3.5$. The reactive power P_{im} starts with a positive value, decreases and reaches 0 for $kw_0 = 1$, decreases further and reaches a minimum value for $kw_0 \approx 1.55$, and then increases and approaches 0 as kw_0 is increased. The paraxial approximation corresponds to $kw_0 \to \infty$. The reactive power does approach the limiting value of 0 of the paraxial beam as kw_0 is increased.

5 Gaussian beam beyond the paraxial approximation

Agrawal and Pattanayak [2] were perhaps the first to obtain a full Gaussian wave corresponding to an input distribution such as that given by Eq. (1.1) that generates the fundamental Gaussian beam. For purposes of comparison, the work of Agrawal and Pattanayak is cast in the framework of the present treatment. Agrawal and Pattanayak considered only the propagation in the $+z$ direction in the space $0 < z < \infty$; here, the propagation in the $-z$ direction in the space $-\infty < z < 0$ is also included. The vector potential of Agrawal and Pattanayak is expressed as

$$A_x^{\pm}(x,y,z) = \frac{N}{ik}\pi w_0^2 \int_{-\infty}^{\infty}\int_{-\infty}^{\infty} dp_x dp_y \exp[-i2\pi(p_x x + p_y y)]$$
$$\times \exp\left[-\pi^2 w_0^2\left(p_x^2 + p_y^2\right)\right]\exp(i\zeta|z|) \tag{37}$$

where ζ is defined in Eq. (3). For $|z| > 0+$ (that is, for $z > 0+$ and $z < 0-$), it can be verified that $A_x^{\pm}(x,y,z)$ given by Eq. (37) satisfies the homogeneous Helmholtz equation:

$$\left(\frac{\partial^2}{\partial x^2} + \frac{\partial^2}{\partial y^2} + \frac{\partial^2}{\partial z^2} + k^2\right)A_x^{\pm}(x,y,z) = 0 \quad \text{for } z \neq 0 \tag{38}$$

For $z = 0$, the integrations in Eq. (37) are carried out to reproduce correctly the input distribution assumed at the outset as given by Eq. (1.1). $A_x^{\pm}(x,y,z)$ given by Eq. (37) is continuous across the plane $z = 0$. The electromagnetic field components $E_x^{\pm}(x,y,z)$ and $H_y^{\pm}(x,y,z)$ are obtained from Eqs. (9), (10), and (37) as

$$E_x^{\pm}(x,y,z) = N\pi w_0^2 \int_{-\infty}^{\infty}\int_{-\infty}^{\infty} dp_x dp_y \exp[-i2\pi(p_x x + p_y y)]$$

$$\times \left(1 - \frac{4\pi^2 p_x^2}{k^2}\right) \exp\left[-\pi^2 w_0^2\left(p_x^2 + p_y^2\right)\right] \exp(i\zeta|z|) \tag{39}$$

and

$$H_y^{\pm}(x,y,z) = \pm\frac{N}{k}\pi w_0^2 \int_{-\infty}^{\infty}\int_{-\infty}^{\infty} dp_x dp_y \exp[-i2\pi(p_x x + p_y y)]$$

$$\times \exp\left[-\pi^2 w_0^2\left(p_x^2 + p_y^2\right)\right]\zeta \exp(i\zeta|z|) \tag{40}$$

$E_x^{\pm}(x,y,z)$ given by Eq. (39) is also continuous across the plane $z = 0$. But $H_y^{\pm}(x,y,z)$ given by Eq. (40) is discontinuous across the plane $z = 0$. The discontinuity of the tangential component of the magnetic field is equivalent to the surface electric current density on the plane $z = 0$. The source electric current density is determined from Eq. (40) as

$$\mathbf{J}(x,y,z) = \hat{z} \times \hat{y}[H_y^+(x,y,0) - H_y^-(x,y,0)]\delta(z)$$

$$= -\hat{x}\frac{2N}{k}\pi w_0^2 \delta(z)\int_{-\infty}^{\infty}\int_{-\infty}^{\infty} dp_x dp_y \exp[-i2\pi(p_x x + p_y y)]$$

$$\times \exp\left[-\pi^2 w_0^2\left(p_x^2 + p_y^2\right)\right]\zeta \tag{41}$$

The source electric current density given by Eq. (41) pertains to the Helmholtz wave equation. The paraxial approximation to Eq. (41) is found by replacing ζ in the amplitude by the first term in Eq. (6). Then the integrals in Eq. (41) are evaluated, with the result that

$$\mathbf{J}_0(x,y,z) = -\hat{x}2N \exp\left[-\frac{x^2 + y^2}{w_0^2}\right]\delta(z) \tag{42}$$

which is identical to that in Eq. (1.24). In the work of Agrawal and Pattanayak, for the exact full wave, the source electric current density is given by Eq. (41), and for the approximate paraxial beam, the source electric current density is given by Eq. (42). Thus, the fundamental Gaussian wave treated here is not the same as the full Gaussian wave introduced by Agrawal and Pattanayak.

References

1. S. R. Seshadri, "Quality of paraxial electromagnetic beams," *Appl. Opt.* **45**, 5335–5345 (2006).

2. G. P. Agrawal and D. N. Pattanayak, "Gaussian beam propagation beyond the paraxial approximation," *J. Opt. Soc. Am.* **69**, 575–578 (1979).

CHAPTER 3

Origin of point current source in complex space

A perfect conductor occupies the $z = 0$ plane. Electromagnetic fields are excited only in $z > 0$ by a point electric dipole situated on the z axis at $z = h$ and oriented in the z direction. The fields are outgoing in $z > 0$, the tangential components of the electric field vanish on $z = 0+$, and there are no sources or fields for $z < 0$. The relevant physical space is $z > 0$, and the rest of the space is external to the physical space. The fields excited by the point electric dipole are determined as follows: At the image location of the real dipole, namely on the z axis at $z = -h$, a virtual dipole of identical strength and oriented in the same direction as the real dipole is placed. The perfect conductor occupying the $z = 0$ plane is removed. Electromagnetic theory is as valid external to the physical space as it is in the physical space. The fields generated by the virtual dipole are determined in the same manner as for the real dipole. The tangential electric fields caused by the real and virtual dipoles on the surface $z = 0+$ are found to vanish identically. A perfect conductor is now inserted on the $z = 0$ plane. The required boundary conditions on $z = 0+$ continue to be satisfied. A perfect conductor is a perfect insulator. The virtual dipole is removed. The fields for $z < 0$ vanish completely. The fields for $z > 0$ are unaffected. All the requirements of the problem are met. The fields are generated by the real dipole for $z > 0$, the fields are outgoing for $z > 0$, the tangential components of the electric field vanish on $z = 0+$, and there are no sources or fields for $z < 0$. Thus, the concept of a virtual source placed external to the relevant physical space is helpful for synthesizing the electromagnetic fields generated by a vertical electric dipole in front of a perfect conductor occupying the $z = 0$ plane (see Fig. 1).

The stated electromagnetic problem can be solved systematically as a boundary-value problem without the need to introduce a virtual source. Several electromagnetic boundary-value problems are solved similarly by the use of virtual sources placed outside the relevant physical space.

The concept of a virtual source, introduced by Deschamps, has some novel features. The usual electromagnetic field theory is not valid in the region where the virtual source is located. The field generated by the virtual source in the complex space reproduces the field of a paraxial scalar fundamental Gaussian beam to within an amplitude term in the real space. This novel virtual point source is now introduced.

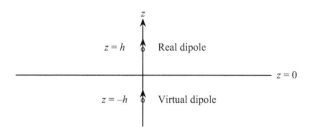

Fig. 1.   Vertical point electric dipole in front of a perfect conductor occupying the plane $z = 0$. Real and virtual dipoles are on the z axis at $z = h$ and $z = -h$, respectively.

1 Scalar Gaussian beam

The wave function, which describes a scalar field, satisfies the Helmholtz equation:

$$\left(\frac{\partial^2}{\partial x^2} + \frac{\partial^2}{\partial y^2} + \frac{\partial^2}{\partial z^2} + k^2 \right) u^{\pm}(x, y, z) = 0 \tag{1}$$

The field that has only wavevectors in a small range of directions about the propagation direction $(\pm z)$ can be expressed as

$$u^{\pm}(x, y, z) = u_p^{\pm}(x, y, z) \exp(\pm ikz) \tag{2}$$

The exponential function is the rapidly varying phase and $u_p^{\pm}(x, y, z)$ is the slowly varying amplitude. The subscript p stands for paraxial. Substituting Eq. (2) into Eq. (1) and simplifying the result as in Eqs. (C1)–(C7) leads to the paraxial wave equation:

$$\left(\frac{\partial^2}{\partial x^2} + \frac{\partial^2}{\partial y^2} \pm 2ik \frac{\partial}{\partial z} \right) u_p^{\pm}(x, y, z) = 0 \tag{3}$$

The field $u^{\pm}(x, y, z)$ at any z is required given the field $u^{\pm}(x, y, 0)$ at the input $z = 0$. Note that $u^{\pm}(x, y, 0) = u_p^{\pm}(x, y, 0)$. The two-dimensional Fourier transform of $u_p^{\pm}(x, y, 0)$ is found from Eq. (A18) as

$$\bar{u}_p^{\pm}(p_x, p_y, 0) = \int_{-\infty}^{\infty} \int_{-\infty}^{\infty} dx_1 dy_1 u_p^{\pm}(x_1, y_1, 0) \exp\left[i2\pi(p_x x_1 + p_y y_1)\right] \tag{4}$$

The Fourier integral representation of $u_p^{\pm}(x, y, z)$ as given by Eq. (A17) is substituted into Eq. (3) to obtain the differential equation for $\bar{u}_p^{\pm}(p_x, p_y, z)$. The solution of the differential equation yields

$$\bar{u}_p^{\pm}(p_x, p_y, z) = \bar{u}_p^{\pm}(p_x, p_y, 0) \exp\left[-i \frac{2\pi^2}{k} \left(p_x^2 + p_y^2 \right) |z| \right] \tag{5}$$

From Eq. (4), $\bar{u}_p^{\pm}(p_x, p_y, 0)$ is substituted into Eq. (5) and the resulting expression for $\bar{u}_p^{\pm}(p_x, p_y, z)$ is inverted, with the result that

$$u_p^\pm(x,y,z) = \int_{-\infty}^\infty \int_{-\infty}^\infty dx_1 dy_1 u_p^\pm(x_1,y_1,0) \int_{-\infty}^\infty \int_{-\infty}^\infty dp_x dp_y \exp\left[-i\frac{2\pi^2}{k}\left(p_x^2 + p_y^2\right)|z|\right]$$
$$\times \exp\{-i2\pi[p_x(x-x_1) + p_y(y-y_1)]\} \tag{6}$$

The inner integrals are evaluated by the use of Eqs. (B1) and (B6) to obtain $u_p^\pm(x,y,z)$. Then, together with Eq. (2), it is found that

$$u^\pm(x,y,z) = -\frac{ik}{2\pi|z|} \exp(\pm ikz) \int_{-\infty}^\infty \int_{-\infty}^\infty dx_1 dy_1 u^\pm(x_1,y_1,0)$$
$$\times \exp\left\{\frac{ik}{2\pi|z|}\left[(x-x_1)^2 + (y-y_1)^2\right]\right\} \tag{7}$$

Equation (7) is the Fresnel integral representation of $u^\pm(x,y,z)$ in terms of the input field distribution $u^\pm(x,y,0)$.

The input value of $u^\pm(x,y,0)$ is assumed as in Eq. (S3), where the prefix S denotes the equations and the references in the paper by Deschamps that is reproduced at the end of this chapter. Here, $\mathbf{x} = \hat{x}x + \hat{y}y$ and $\mathbf{x}^2 = x^2 + y^2$. The variance of the input field distribution is $2a/k = 2\sigma_0^2$. Given the input field distribution, there are two ways to proceed: First, $\bar{u}_p^\pm(p_x,p_y,0)$ is found from Eq. (4), and $\bar{u}_p^\pm(p_x,p_y,z)$ is obtained from Eq. (5). Then, $\bar{u}_p^\pm(p_x,p_y,z)$ is inverted to determine $u_p^\pm(x,y,z)$. The rapidly varying phase is inserted in accordance with Eq. (2) to find $u^\pm(x,y,z)$. This procedure is followed for deducing the vector potential in Section 1.1. Alternatively, the input field distribution $u^\pm(x,y,0)$ is substituted directly into the Fresnel integral representation given by Eq. (7) and the integrals are evaluated by the use of Eqs. (B1) and (B6). The result is

$$u^\pm(\mathbf{x},z) = \left(1 + i\frac{|z|}{a}\right)^{-1} \exp(ik|z|) \exp\left(-\frac{k\mathbf{x}^2}{2A}\right) \exp\left(i\frac{k\mathbf{x}^2}{2R}\right) \tag{8}$$

where A is given by Eq. (S5) and

$$R = |z| + \frac{a^2}{|z|} \tag{9}$$

Deschamps provided the details only for propagation in the $+z$ direction. Here propagation in the $\pm z$ direction is considered. The upper and the lower signs correspond to propagation in the $\pm z$ direction.

In Eq. (8) [same as Eq. (S4)], the first two terms show the variation of the field $u^\pm(0,z)$ along the axis. The third term reveals the spreading of the beam in the transverse (x,y) directions. The variance of the cross-sectional distribution is $2A/k$, which increases quadratically with z in accordance with Eq. (S5). The fourth term gives the radius of curvature of the phase front. The radius of curvature is infinite for $|z| = 0$ and $|z| = \infty$; it has the minimum value $R_{\min} = 2a$ for $|z| = a$. The two real numbers (A,R), which describe the cross section at z of the beam, can be combined into a single complex number Z, as shown in Eq. (S7). The complex number Z is defined by

$$Z = |z| - ia \tag{10}$$

Then, $u^{\pm}(\mathbf{x}, z)$ given by Eq. (8) can be expressed as

$$u^{\pm}(\mathbf{x}, z) = -\frac{ia}{Z} \exp(ik|z|) \exp\left(i\frac{k\mathbf{x}^2}{2Z}\right) \tag{11}$$

A simple explanation of this result was proposed by Deschamps at the 1967 meeting of the Union Radio-Scientifique Internationale (URSI) [S9] and later at the 1968 URSI Symposium on Electromagnetic Waves [S10]. The beam representation, on which the explanation was based, was presented in 1971 in a letter that is reproduced at the end of this chapter.

2 Field of a point source

Deschamps provided an important explanation of the result obtained in Eq. (11). A point source of strength 4π capable of generating a scalar field satisfies the following inhomogeneous Helmholtz equation:

$$\left(\frac{\partial^2}{\partial x^2} + \frac{\partial^2}{\partial y^2} + \frac{\partial^2}{\partial z^2} + k^2\right) G(x, y, z) = -4\pi\delta(x)\delta(y)\delta(z) \tag{12}$$

The source is situated at the origin, $x = y = z = 0$. It should be remembered that x, y, and z are real. As is derived in Appendix A, the solution of Eq. (12) is

$$G(\mathbf{x}, z) = \frac{\exp(ikr)}{r} \text{ with } r = (x^2 + y^2 + z^2)^{1/2} \tag{13}$$

If r is expanded for $x^2 + y^2 \ll z^2$, it is found that

$$r = |z| + \frac{(x^2 + y^2)}{2|z|} \tag{14}$$

Substituting Eq. (14) into Eq. (13), keeping only the first term in the amplitude and only the first two terms in the phase yields

$$G(\mathbf{x}, z) = \frac{1}{|z|} \exp(ik|z|) \exp\left(i\frac{k\mathbf{x}^2}{2|z|}\right) \tag{15}$$

By keeping the second term of the expansion also in the amplitude, a term that is one more order of magnitude lower than that stated in Eq. (15) is obtained; therefore, it is not retained. If $|z|$ in Eq. (15) is changed to Z, Eq. (11) becomes equal to Eq. (15), except that Eq. (11) has an extra factor: $-ia \exp(-ka)$. For $|z| = 0$, $|Z| = a = \frac{1}{2}k(\sqrt{2}\sigma_0)^2$. The Rayleigh distance $a \gg \sqrt{2}\sigma_0$, where $\sqrt{2}\sigma_0$ is the beam waist. Usually the beam waist is significantly larger than the wavelength, that is, $\sqrt{2}\sigma_0 \gg \lambda = \frac{2\pi}{k}$; therefore, $\sqrt{2}\sigma_0 k \gg 1$. Thus, the relation between Eq. (15) and Eq. (11) is valid for all z, including $z = 0$. There is no electromagnetic theory when the position coordinates are complex with nonvanishing imaginary parts. Therefore, the electromagnetic fields of a point source located at $x = 0$, $y = 0$, and $|z| = ia$ cannot be found directly from the application of the present electromagnetic theory.

Yet Deschamps made a daring departure from convention to postulate that the field of a virtual point source approximates correctly the paraxial beam field to within an amplitude factor in the paraxial approximation: $k\sqrt{2}\sigma_0 \gg 1$. However, a systematic procedure exists for deducing the field in the real space caused by the virtual point source located in the complex space where one of the position coordinates of the point source is imaginary. The field is first determined for real z and then continued analytically from real z to complex $(|z| - ia)$. The full implementation of this procedure is discussed in Chapter 5.

3 Extensions

Deschamps considered propagation only in the $+z$ direction. For propagation in the $+z$ direction, the location of the virtual source is at $x = 0$, $y = 0$, and $z = ia$. For propagation in the $-z$ direction, the location of the virtual source is at $x = 0$, $y = 0$, and $|z| = ia$ or $z = -ia$. The location of the virtual source is different for the two opposite directions of propagation. It should also be noted that the Rayleigh distance a is usually very large compared to the waist size $\sqrt{2}\sigma_0$ of the beam at the input.

Deschamps treated only the scalar field. The treatment can be extended to the vector fields. In general, the electromagnetic fields can be determined in terms of a single component of the magnetic vector potential and a single component of the electric vector potential, both in the same direction. The paraxial approximation for the two potentials is found in the same manner as that for the scalar field. The electromagnetic fields are then determined by differential operations on the two potentials. These operations introduce terms that are of higher order than the potentials. The higher-order terms are omitted to obtain the paraxial approximation for the electromagnetic fields. The transverse derivatives (x, y) act only on the slowly varying amplitude, and the resulting terms are one order of magnitude in $1/k\sqrt{2}\sigma_0$ smaller than the potentials. The longitudinal (z) derivative acting on the rapidly varying phase is taken to be the same order of magnitude as the potentials. Then, the longitudinal (z) derivative acting on the slowly varying amplitude yields terms that are two orders of magnitude in $1/k\sqrt{2}\sigma_0$ smaller than the potentials. Only the terms that are appropriate for the paraxial approximation are retained. Then, the paraxial electromagnetic Gaussian beam is obtained.

Deschamps treated only the fields that are far from the source, that is, the radiation fields. For electric current sources, the fields near the source, and even at the source, are important [1, 2]. In other words, the input properties of the current sources are also important in order to find out the mechanisms of excitation of the current source. For point current sources, the reactive power is infinite. Physically meaningful finite values of reactive power can be obtained only by taking into account the finite dimensions of the sources. Therefore, there is a need to extend the work of Deschamps by including the finite dimensions of the source.

The analyses of resonator modes by Fox and Li [S2], Boyd and Gordon [3], Boyd and Kogelnik [S3], Kogelnik and Li [4], and Arnaud and Kogelnik [S7] lead to the scalar fundamental Gaussian beam as described by Eq. (8). The field $G(\mathbf{x}, Z)$ as given by Eq. (15) can differ from $u^{\pm}(\mathbf{x}, z)$ as given by Eq. (11) outside the paraxial region $k\sqrt{2}\sigma_0 \gg 1$. For example, the Green's function $G(\mathbf{x}, z)$ given by Eq. (13), with z replaced by Z as given by Eq. (10), is singular for $r = 0$. This singularity occurs on a circle $x^2 + y^2 = a^2$ with the center at $z = 0$ and axis along the z axis. The circle is a branch line for the function $r(\mathbf{x}, Z)$. The plane circular sheet defined by $z = 0$ and $|\mathbf{x}| > a$ is used as a branch cut; then, the branch of $G(\mathbf{x}, Z)$ that approximates $u^{\pm}(\mathbf{x}, z)$ as given by Eq. (11) represents the scalar fundamental Gaussian beam that propagates in the $\pm z$ direction and decreases as $|\mathbf{x}|$ increases.

4 Exact solution

The fundamental Gaussian beam is the solution of the parabolic equation, that is, the approximate paraxial wave equation. The fundamental Gaussian wave is the solution of the exact Helmholtz equation. Therefore, the fundamental Gaussian wave is expected to be a better representation of the field than the fundamental Gaussian beam. Deschamps also indicated a method of obtaining the exact solution of the Helmholtz equation. The Green's function $G(\mathbf{x}, Z)$ is an exact solution of the Helmholtz equation. By setting

$$r = \left[\mathbf{x}^2 + (|z| - ia)^2 \right]^{1/2} = r' + ir''$$

the curves on which r' and r'' are constant form a system of confocal ellipses and hyperbolas. But Deschamps did not suggest a specific wave function. The suggestion of Deschamps has been implemented, for example, by Landesman and Barrett [5], who have determined the exact solution of the Helmholtz equation in the prolate and the oblate spheriodal coordinate systems.

References

1. S. R. Seshadri, "Constituents of power of an electric dipole of finite size," *J. Opt. Soc. Am. A* **25**, 805–810 (2008).

2. S. R. Seshadri, "Power of a simple electric multipole of finite size," *J. Opt. Soc. Am. A* **25**, 1420–1425 (2008).

3. G. D. Boyd and J. P. Gordon, "Confocal multimode resonator for millimeter through optical wavelength masers," *Bell Syst. Tech. J.* **40**, 489–508 (1961).

4. H. Kogelnik and T. Li, "Laser beams and resonators," *Appl. Opt.* **5**, 1550–1567 (1966); *Proc. IEEE* **54**, 1312–1329 (1966).

5. B. T. Landesman and H. H. Barrett, "Gaussian amplitude functions that are exact solutions to the scalar Helmholtz equation," *J. Opt. Soc. Am. A* **5**, 1610–1619 (1988).

Reproduced from *Electronics Letters*, Volume 7, Issue 23,
18 November 1971, pp. 684–685

GAUSSIAN BEAM AS A BUNDLE OF COMPLEX RAYS

Indexing terms: *Optical propagation effects, Electromagnetic-wave scattering, Electromagnetic-wave diffraction*

It is observed that the function $G(P) = e^{ikr}/r$, where r is the distance of the observation point P to a fixed point having a complex location, represents the field of a Gaussian beam. This can be used to justify, without further computation, the application of formulas of ordinary optics to the transformation of beams through optical systems. It can also be used to solve very simply some problems of diffraction and scattering of Gaussian beams.

A beam of electromagnetic radiation is loosely defined as a field which propagates substantially close to a central line, which may be called its axis. Beams have been intensely studied during the past ten years because of their applications to the guiding of electromagnetic energy (Goubau beam waveguide[1]), and to coherent optics (lasers and resonators).

The paraxial method of analysis takes advantage of the fact that the angular spectrum of a beam is made up of plane waves whose wavevectors are close to the axis. The wavenumber ζ in the direction of the axis z may then be expressed in terms of the transverse wavevector $\boldsymbol{\kappa}$ by

$$\zeta \sim k + \frac{1}{2k} \cdot \boldsymbol{\kappa}^2 \qquad (1)$$

This may be considered as approximating the dispersion surface, a sphere of radius k, by a paraboloid of the same curvature (eqn. 1) at $\boldsymbol{\kappa} = 0$. To this approximation corresponds an approximation of the wave equation $(\Delta + k^2)u = 0$ by the parabolic equation

$$\frac{1}{i} \frac{\partial}{\partial z} = k - \frac{1}{2k} \left(\frac{\partial^2}{\partial x^2} + \frac{\partial^2}{\partial y^2} \right) \qquad (2)$$

Either of eqns. 1 or 2 may be used to determine how the field in a plane transverse to the beam evolves with the change in z. If this evolution is such that the field remains constant except for a change in scale and a change in the phase-front curvature, the field constitutes a beam mode. These modes have been extensively studied. References 2–7 represent only a small sample of this work.

If all the sources are in the region $z < 0$ and if the field at $z = 0$ is

$$u(\mathbf{x}, 0) = \exp\left(-\frac{1}{2} \frac{\mathbf{x}^2}{\sigma_0^2} \right) = \exp\left(-\frac{k}{2} \frac{\mathbf{x}^2}{a} \right) \qquad (3)$$

where $a = k\sigma_0^2$ is the scaled variance and $\mathbf{x}$ is the transverse position vector, a Gaussian beam is generated in the region $z > 0$. Using the paraxial approximation, it is found that

$$u(\mathbf{x},z) = \left(1 + i\frac{z}{a}\right)^{-1} e^{ikz} \exp\left(-\frac{1}{2}\frac{\mathbf{x}^2}{\sigma^2}\right) \exp\left(\frac{ik}{2}\frac{\mathbf{x}^2}{R}\right) \tag{4}$$

The first two terms indicate the variation of the field $u(0,z)$ along the axis. The next describes the spread of the beam. It may be written $\exp\{-k\mathbf{x}^2/(2A)\}$, with a scaled variance $A = k\sigma^2$ which increases with z according to

$$A = a + \frac{z^2}{a} \tag{5}$$

The last term shows the curvature of the phase front. The radius is

$$R = z + \frac{a^2}{z} \tag{6}$$

It has been noted[4,6] that the two real numbers (A, R), which characterize the beam cross-section at z, can be combined into a single complex number Z defined by

$$\frac{1}{Z} = \frac{1}{R} + \frac{i}{A} \tag{7}$$

Eqns. 5 and 6 then reduce to

$$Z = z - ia \tag{8}$$

And eqn. 4 becomes

$$u(\mathbf{x},z) = -\frac{ia}{Z} e^{ikz} \exp\left(\frac{ik}{2}\frac{\mathbf{x}^2}{Z}\right) \tag{9}$$

Furthermore, if a Gaussian beam propagates along the axis of a symmetric optical system made up of coaxial lenses and mirrors, the transformation between Z at the input 0 to the system, and Z' at its output $0'$, is of the form

$$Z' = (aZ + b)(cZ + d)^{-1} \tag{10}$$

It was observed in eqn. 4 that this formula is identical to that which relates in optics the abscissa Z (with respect to 0) of a point source and the abscissa Z' (with respect to $0'$) of its image. There Z and Z' are real, but the coefficients (a, b, c, d) are the same as in eqn. 10. Thus the evolution of a beam through an optical system is easily deduced from the usual lens and mirror formulas.

A simple explanation of this result was proposed by the author at the 1967 fall URSI meeting and later at the Stresa International symposium of URSI.[9,10] The purpose of this letter is to discuss the beam representation on which the explanation was based and to suggest further applications of this representation.

The function $G(\mathbf{x}, z) = e^{ikr}/r$, where r is the distance of the point $(\mathbf{x}, z)$ to some fixed origin, is a solution of the wave equation. If we choose for the origin the point $C(\mathbf{x} = 0, z = ia)$, then r becomes complex and the function G, near to the z axis, represents a Gaussian beam. This is easily verified. Since $|\mathbf{x}| \ll z - ia$,

$$r = \left\{ (z - ia)^2 + \mathbf{x}^2 \right\}^{1/2} \sim z - ia + \frac{1}{2} \frac{\mathbf{x}^2}{z - ia}$$

And therefore

$$G(\mathbf{x}, z) \sim \frac{e^{ik(z-ia)}}{z - ia} \exp\left(\frac{ik}{2} \frac{\mathbf{x}^2}{z - ia} \right) \tag{11}$$

(In the denominator of G, $r \sim z - ia$ as usual.) We recognize the dependence on x, which is identical to that in the beam of eqn. 9. $Z = z - ia$ is the distance to the complex centre C. Along the axis, $G(0, z)$ differs from $u(0, z)$ only by a constant factor $-e^{ka}/ia$. Thus eqn. 11 accounts not only for the transverse variation of the field but also for its z dependence, including the phase correction $\tan^{-1}(a/z)$ to kz which comes from the complex amplitude.

The Gaussian beam is therefore equivalent paraxially to a spherical wave with a centre C at a complex location. This may be used to solve problems of refraction, diffraction and scattering of beams, when the corresponding solution for the central field G is known. The idea has been applied to a few simple problems, namely diffraction by a straight edge and scattering by a cylinder.[11] One must, of course, pay attention to the regions of validity of the identification. The field G outside the paraxial region can differ appreciably from the Gaussian beam. In particular, G is singular when $r = 0$, which occurs on a circle of radius a with center o and axis oz. This circle is also a branch line for the function $r(\mathbf{x}, z)$. Using as a branch cut the surface $(z = 0, |\mathbf{x}| > a)$, a branch of G that approximates eqn. 11 represents a field that propagates towards $z > 0$ and decreases as $|\mathbf{x}|$ increases. Instead of eqn. 11, one may use the exact expression for G with $r = r' + ir''$. The curves on which r' and r'' are constant form a system of confocal ellipses and hyperbolas. This exact G satisfies the wave equation instead of the parabolic equation and should therefore be a better representation of the field.

The spherical wave G can also be interpreted as a bundle of complex rays originating from C. More generally, bundles of complex rays can be handled by geometrical optic methods[7,10] and they represent general Gaussian beams where the field in a cross-section is of the form $e^{i\phi(\mathbf{x})}$ with a complex phase

$$\phi(\mathbf{x}) = \frac{1}{2} k (\mathbf{x}^T Z^{-1} \mathbf{x})$$

The transformation of the matrix Z is still given by eqn. 10. For a nonsymmetrical optical system, eqn. 11 is still valid, but (a, b, c, d) become 2×2 matrices.[6,7,10]

Gaussian beams have also been analysed in terms of complex rays by Keller and Streifer.[8] The construction they use would apply to more general cases of beam propagation. However, the formula they derive from it is only valid when $z \gg a$, while our expression is valid even for $z = 0$ if $|\mathbf{x}| \ll a$.

G. A. Deschamps
Antenna Laboratory
University of Illinois
Urbana, Ill. 61801, USA

11th October 1971

References

1. GOUBAU, G., and SCHWERING, F.: 'On the guided propagation of electromagnetic wave beams', *IRE Trans.*, 1961, **AP-9**, pp. 248–256

2. FOX, A. G., and LI, T.: 'Resonant modes in a maser interferometer', *Bell Syst. Tech. J.*, 1961, **40**, pp. 453–488

3. BOYD, G. D., and KOGELNIK, H.: 'Generalized confocal resonator theory', *ibid.*, 1962, **41**, pp. 1347–1369

4. DESCHAMPS, G. A., and MAST, P. E.: 'Beam tracing and applications' *in* 'Proceedings of the symposium on quasioptics' (Polytechnic Press, New York, 1964), pp. 379–395

5. COLLINS, S. A.: 'Analysis of optical resonators involving focusing elements', *Appl. Opt.*, 1964, **3**, pp. 1263–1275

6. KOGELNIK, H.: 'On the propagation of Gaussian beams of light through lenslike media including those with a loss or gain variation', *ibid.*, 1965, **4-12**, pp. 1562–1569

7. ARNAUD, J. A., and KOGELNIK, H.: 'Gaussian light-beams with general astigmatism', *ibid.*, 1969, **8**, pp. 1687–1693

8. KELLER, JOSEPH B., and STREIFFER, W.: 'Complex rays with an application to Gaussian beams', *J. Opt. Soc. Am.*, 1971, **61-1**, pp. 40–43

9. DESCHAMPS, G. A.: 'Matrix methods in geometrical optics', 1967 fall meeting of URSI, University of Michigan, p. 84 of abstract

10. DESCHAMPS, G. A.: 'Beam optics and complex rays', URSI symposium on electromagnetic waves, Stresa, June 1968

11. GOWAN, E., and DESCHAMPS, G. A.: 'Quasi-optical approaches to the diffraction and scattering of Gaussian beams', University of Illinois Antenna Laboratory report 70-5, Urbana-Champaign, Ill., USA

CHAPTER 4

Basic full Gaussian wave

The field of a point source of suitable strength with the location coordinates in the complex space reproduces the fundamental Gaussian beam in the paraxial approximation. The resulting full wave is designated the basic full Gaussian wave. The vector potential that generates the basic full Gaussian wave is obtained. The resulting electromagnetic fields are found, the radiation intensity distribution is determined, and its characteristics are described. The time-averaged power transported by the basic full Gaussian wave in the $+z$ and the $-z$ directions is obtained. This power increases monotonically and approaches the limiting value of the fundamental Gaussian beam as kw_0 is increased. The surface electric current density on the secondary source plane $z = 0$ is deduced. In the paraxial approximation, this electric current density reduces to the source electric current density that generates the fundamental Gaussian beam. In the paraxial approximation, both the fundamental Gaussian wave and the basic full Gaussian wave reduce to the same fundamental Gaussian beam. The complex power is evaluated and the reactive power is determined. For the basic full Gaussian wave, the reactive power is infinite. In contrast, for the corresponding paraxial beam, namely the fundamental Gaussian beam, the reactive power vanishes.

1 Point source in complex space

Deschamps [1] postulated that the paraxial approximation of the field due to a point source reproduced the fundamental Gaussian beam provided that the location coordinates of the source are in the complex space. Felsen [2, 3] indicated analytical continuation as the tool for the interpretation of the field due to a point source with the location coordinates in the complex space with reference to the field due to the same source with the location coordinates in the real space. Both Deschamps and Felsen considered only the outward propagation in the $+z$ direction in the physical space $0 < z < \infty$. In this development, the outward propagation in the $-z$ direction in the physical space $-\infty < z < 0$ is also included. Only a scalar point source is considered, and the resulting field is assumed to be the x component of the magnetic vector potential $A_x^{\pm}(x, y, z)$. The field due to the point source at $x = 0$, $y = 0$, and $z = 0$ is analytically continued to yield the field due to the point source at $x = 0$, $y = 0$, and $|z| = ib$, where $b = \frac{1}{2}kw_0^2$ is the Rayleigh distance. The strength of the source is assumed as $(N/ik)\pi w_0^2 S_{ex}$, where S_{ex} is the excitation coefficient and $(N/ik)\pi w_0^2$ is introduced for later convenience. From Eqs. (A15) and (A16), the analytically continued field is found as

$$A_x^{\pm}(x,y,z) = \frac{N}{ik}\pi w_0^2 S_{ex} \frac{\exp\left\{ik\left[x^2+y^2+(|z|-ib)^2\right]^{1/2}\right\}}{4\pi\left[x^2+y^2+(|z|-ib)^2\right]^{1/2}} \tag{1}$$

where $\pm$ denotes the propagation in the $\pm z$ direction. When $(|z|-ib)^2$ is taken outside the square root operation, the branch with the positive real part $(|z|-ib)$ is chosen. In the paraxial approximation, $(x^2+y^2)/|(|z|-ib)^2| \ll 1$, Eq. (1) simplifies as

$$A_x^{\pm}(x,y,z) = \frac{N}{ik}\pi w_0^2 S_{ex} \frac{q_{\pm}^2}{(-4\pi ib)}\exp(\pm ikz)$$

$$\times \exp(kb)\exp\left[-\frac{q_{\pm}^2(x^2+y^2)}{w_0^2}\right] \tag{2}$$

In obtaining the paraxial approximation, $[x^2+y^2+(|z|-ib)^2]^{1/2}$ is expanded into a power series in the small parameter $(x^2+y^2)/|(|z|-ib)^2|$; the first term is retained in the amplitude and the first two terms are retained in the phase. The excitation coefficient is chosen as

$$S_{ex} = -2ik\exp(-kb) \tag{3}$$

Then Eq. (2) reduces to

$$A_x^{\pm}(x,y,z) = \exp(\pm ikz)\frac{N}{ik}q_{\pm}^2\exp\left[-\frac{q_{\pm}^2(x^2+y^2)}{w_0^2}\right] \tag{4}$$

which is identical to the expression given by Eq. (1.11) for the vector potential that generates the fundamental Gaussian beam. In Eq. (1), S_{ex} from Eq. (3) is substituted. Then, the full-wave solution given by Eq. (1) changes as

$$A_x^{\pm}(x,y,z) = \frac{N}{ik}\left[-\pi w_0^2 2ik\exp(-kb)\right]$$

$$\times \frac{\exp\left\{ik\left[x^2+y^2+(|z|-ib)^2\right]^{1/2}\right\}}{4\pi\left[x^2+y^2+(|z|-ib)^2\right]^{1/2}} \tag{5}$$

In the paraxial approximation, $A_x^{\pm}(x,y,z)$ reduces correctly to $A_{x0}^{\pm}(x,y,z)$, given by Eq. (1.11), where $A_{x0}^{\pm}(x,y,z)$ is the x component of the vector potential that generates the fundamental Gaussian beam in $|z| > 0$. Outside the source region, $A_x^{\pm}(x,y,z)$ satisfies the homogeneous Helmholtz equation. Therefore, $A_x^{\pm}(x,y,z)$ satisfies the homogeneous Helmholtz equation in the physical space $|z| > 0$. There is no implication that, in the plane $z = 0$ that forms the boundary between the physical space and the complex space, $A_x^{\pm}(x,y,z)$ satisfies the homogeneous Helmholtz equation.

Certain differential operations have to be performed on $A_x^{\pm}(x,y,z)$ to derive the associated electromagnetic field components. These differentiations are carried out conveniently if

$A_x^{\pm}(x,y,z)$ is expressed as a two-dimensional inverse Fourier transform. For the point source of unit strength without the stated analytical continuation, the required integral expression is the inverse Fourier transform, as given by Eq. (A26). When the strength of the source and the analytical continuation are included, $A_x^{\pm}(x,y,z)$ given by Eq. (5) is expressed as

$$A_x^{\pm}(x,y,z) = \frac{N}{i}\pi w_0^2 \exp(-kb) \int_{-\infty}^{\infty}\int_{-\infty}^{\infty} dp_x dp_y$$

$$\times \exp[-i2\pi(p_x x + p_y y)]\zeta^{-1}\exp[i\zeta(|z| - ib)] \tag{6}$$

where ζ is defined by Eq. (2.3). The full-wave generalization of the fundamental Gaussian beam as introduced by Deschamps and Felsen is designated the basic full Gaussian wave.

2 Electromagnetic fields

The associated electromagnetic fields are obtained by substituting $A_x^{\pm}(x,y,z)$ from Eq. (6) into Eqs. (D30)–(D33). For the determination of the Poynting vector and the complex power, only the field components $E_x^{\pm}(x,y,z)$ and $H_y^{\pm}(x,y,z)$ are required. From Eqs. (2.9), (2.10), and (6), it is found that

$$E_x^{\pm}(x,y,z) = N\pi k w_0^2 \exp(-kb) \int_{-\infty}^{\infty}\int_{-\infty}^{\infty} dp_x dp_y$$

$$\times \left(1 - \frac{4\pi^2 p_x^2}{k^2}\right)\exp[-i2\pi(p_x x + p_y y)]$$

$$\times \zeta^{-1}\exp[i\zeta(|z| - ib)] \tag{7}$$

and

$$H_y^{\pm}(x,y,z) = \pm N\pi w_0^2 \exp(-kb) \int_{-\infty}^{\infty}\int_{-\infty}^{\infty} d\bar{p}_x d\bar{p}_y$$

$$\times \exp[-i2\pi(\bar{p}_x x + \bar{p}_y y)]\exp[i\bar{\zeta}(|z| - ib)] \tag{8}$$

where $\bar{\zeta}$ is the same as ζ with p_x and p_y replaced by $\bar{p}_x$ and $\bar{p}_y$, respectively. The time-averaged power flow per unit area in the $\pm z$ direction, $S_z^{\pm}(x,y,z)$, is determined by substituting Eqs. (7) and (8) into Eq. (1.13). The time-averaged power $P_b^{\pm}$ transported by the basic full Gaussian wave in the $\pm z$ direction is found by integrating $[\pm S_z^{\pm}(x,y,z)]$ with respect to x and y across the entire transverse plane. The expression for $P_b^{\pm}$ is simplified by the use of a procedure similar to that employed in obtaining Eq. (2.15), with the result that

$$P_b^{\pm} = \frac{c}{2}N^2\pi^2 k w_0^4 \exp(-2kb)\,\text{Re}\int_{-\infty}^{\infty}\int_{-\infty}^{\infty} dp_x dp_y$$

$$\times \left(1 - \frac{4\pi^2 p_x^2}{k^2}\right)\zeta^{-1}\exp[i|z|(\zeta - \zeta^*)]\exp[b(\zeta + \zeta^*)] \tag{9}$$

The value of N^2 is substituted from Eq. (1.2). Then, $P_b^\pm$ in the limit of the paraxial approximation equals $P_0^\pm = 1$ W. Equation (9) is simplified in the same manner as in obtaining Eqs. (2.21), (2.22), (2.25), and (2.26) from Eq. (2.15) to yield [4]

$$P_b^\pm = \int_0^{\pi/2} d\theta^\pm \sin\theta^\pm \int_0^{2\pi} d\phi\, \Phi(\theta^\pm, \phi) \tag{10}$$

where

$$\Phi(\theta^\pm, \phi) = \frac{\left(1 - \sin^2\theta^\pm \cos^2\phi\right)\exp\left[-k^2 w_0^2\left(1 - \cos\theta^\pm\right)\right]}{2\pi f_0^2} \tag{11}$$

And, as defined by Eq. (2.24),

$$\theta^- = \pi - \theta^+ \tag{12}$$

The basic full Gaussian wave is seen from Eq. (11) to be characterized by one wave parameter kw_0. Equation (11) is substituted into Eq. (10), and the ϕ integration is carried out to obtain

$$P_b^\pm = \frac{1}{f_0^2} \int_0^{\pi/2} d\theta^\pm \sin\theta^\pm \left(1 - \frac{1}{2}\sin^2\theta^\pm\right)$$
$$\times \exp[-k^2 w_0^2 (1 - \cos\theta^\pm)] \tag{13}$$

The variable of integration is changed as

$$1 - \cos\theta^\pm = f_0^2 t = (kw_0)^{-2} t \tag{14}$$

Then Eq. (13) simplifies as

$$P_b^\pm = \int_0^{k^2 w_0^2} dt \left(1 - f_0^2 t + \frac{1}{2} f_0^4 t^2\right) \exp(-t) \tag{15}$$

By repeated integration by parts, $P_b^\pm$ is found as [4]

$$P_b^\pm = 1 - f_0^2 + f_0^4 - \left(\frac{1}{2} + f_0^4\right)\exp\left(-k^2 w_0^2\right) \tag{16}$$

This expression is valuable for evaluating $P_b^\pm$ for $f_0^2 \ll 1$, that is, in the paraxial approximation or the high-frequency region. Equation (16) is exact and can be used for determining $P_b^\pm$ even in the low-frequency region, that is, for $k^2 w_0^2 \ll 1$. The low-frequency behavior is not clear from the form of Eq. (16). If $\exp(-k^2 w_0^2)$ is expanded into a power series and the terms up to $k^6 w_0^6$ are retained, the leading term in $P_b^\pm$ in powers of $k^2 w_0^2$ is given by

$$P_b^\pm = \frac{2}{3} k^2 w_0^2 \tag{17}$$

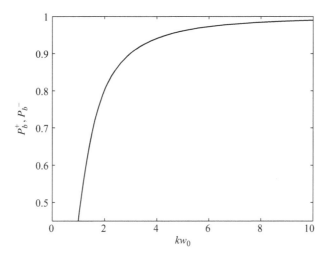

Fig. 1. Time-averaged power $P_b^\pm$ transported by the basic full Gaussian wave in the $\pm z$ direction as a function of kw_0 for $1 < kw_0 < 10$. Time-averaged power transported by the corresponding fundamental Gaussian beam is $P_0^\pm = 1$ W.

Therefore, in the low frequencies, $P_b^\pm$ increases as $2k^2 w_0^2/3$ as kw_0 is increased. In Fig. 1, $P_b^\pm$ is shown as a function of kw_0 for $1 < kw_0 < 10$. $P_b^\pm$ increases monotonically and approaches the limiting value of $P_0^\pm = 1$ W corresponding to the fundamental Gaussian beam as kw_0 is increased.

The ratio of the time-averaged power $P_0^\pm$ in the fundamental Gaussian beam to the time-averaged power $P_b^\pm$ in the basic full Gaussian wave is valuable for determining the quality of the paraxial beam approximation to the exact full wave for different values of the wave parameter kw_0 [5]. Since $P_b^\pm/P_0^\pm$ approaches 1 as kw_0 is increased, the quality of the paraxial beam approximation to the full wave improves as kw_0 is increased.

3 Radiation intensity

The radiation intensity is the time-averaged power flow per unit solid angle in the specified direction. Therefore, it follows from Eqs. (10) and (11) that $\Phi(\theta^+,\phi)$ and $\Phi(\theta^-,\phi)$ are the radiation intensities for $z > 0$ and $z < 0$, respectively. For $z > 0$, θ^+ is defined with respect to the $+z$ axis, and for $z < 0$, θ^- is defined with respect to the $-z$ axis. In view of Eq. (12), it follows that $\Phi(\theta^-,\phi)$ is the mirror image of $\Phi(\theta^+,\phi)$ about the $z = 0$ plane. Therefore, it is sufficient to describe only the radiation intensity distribution for $z > 0$ [4].

The radiation intensity $\Phi(\theta^+,\phi)$ is a function of ϕ, but it has reflection symmetries about the $\phi = (0°, 180°)$ and $\phi = (90°, 270°)$ planes; therefore, the variation of $\Phi(\theta^+,\phi)$ for one quadrant only, namely for $0° < \phi < 90°$, is discussed. In every azimuthal plane $\phi = $ constant, $\Phi(\theta^+,\phi)$ has the value $1/2\pi f_0^2$ for $\theta^+ = 0°$, decreases monotonically as θ^+ increases, and reaches a minimum value for $\theta^+ = 90°$. This minimum value increases from 0 for $\phi = 0°$ to a maximum value for $\phi = 90°$. The beam width of the basic full Gaussian wave has a minimum value for $\phi = 0°$, increases continuously as ϕ is increased, and reaches a maximum value for $\phi = 90°$.

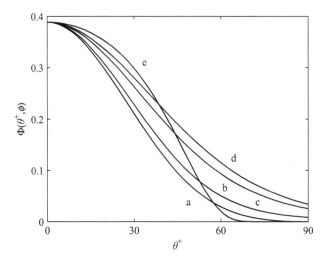

Fig. 2. Radiation intensity pattern $\Phi(\theta^+, \phi)$ of the basic full Gaussian wave for curves (a) $\phi = 0°$, (b) $\phi = 30°$, (c) $\phi = 60°$, (d) $\phi = 90°$, and (e) cylindrically symmetric radiation intensity pattern, $\Phi_0(\theta^+, \phi)$ of the corresponding fundamental Gaussian beam as functions of θ^+ for $0° < \theta^+ < 90°$. Other parameters are $kw_0 = 1.563$; the total power in the fundamental Gaussian beam is 2 W, and that in the basic full Gaussian wave is 1.4000 W.

In Fig. 2, the radiation intensity pattern $\Phi(\theta^+, \phi)$ of the basic full Gaussian wave is shown as functions of θ^+ for $0° < \theta^+ < 90°$, $kw_0 = 1.563$, and for $\phi = 0°$, $30°$, $60°$, and $90°$. The radiation intensity pattern $\Phi_0(\theta^+, \phi)$ of the corresponding fundamental Gaussian beam is included for purposes of comparison. $\Phi_0(\theta^+, \phi)$ is independent of ϕ. The normalization is such that the time-averaged power $P_0^{\pm}$ in the fundamental Gaussian beam propagating in the $+z$ or the $-z$ direction is 1 W. $\Phi_0(\theta^+, \phi)$ is a good approximation to $\Phi(\theta^+, \phi)$ for all ϕ and for $0° < \theta^+ < 90°$. The quality of the agreement between $\Phi(\theta^+, \phi)$ and $\Phi_0(\theta^+, \phi)$ increases as ϕ is increased from $0°$ to $90°$. Therefore, the overall differences between $\Phi(\theta^+, \phi)$ and $\Phi_0(\theta^+, \phi)$ are largest for $\phi = 0°$.

A comparison of Fig. 2 with Fig. 2.2 shows that the essential features of the radiation intensity patterns of the basic full Gaussian wave and the fundamental Gaussian wave are nearly the same.

4 Radiative and reactive powers

$A_x^{\pm}(x, y, z)$, given by Eq. (6), and $E_x^{\pm}(x, y, z)$, given by Eq. (7), are continuous across the $z = 0$ plane. But, $H_y^{\pm}(x, y, z)$ given by Eq. (8), has a discontinuity across the $z = 0$ plane. The discontinuity of the tangential component of the magnetic field is equivalent to a surface electric current density on the plane $z = 0$. The surface electric current density is found from Eq. (8) as

$$\mathbf{J}(x, y, z) = \hat{z} \times \hat{y}[H_y^+(x, y, 0) - H_y^-(x, y, 0)]\delta(z)$$

$$= -\hat{x}2N\pi w_0^2 \exp(-kb)\delta(z)\int_{-\infty}^{\infty}\int_{-\infty}^{\infty} d\overline{p}_x d\overline{p}_y$$

$$\times \exp[-i2\pi(\overline{p}_x x + \overline{p}_y y)]\exp(\overline{\zeta}b) \tag{18}$$

The complex power is determined by substituting Eqs. (7) and (18) in Eq. (D18) as

$$P_C = -\frac{c}{2} \int_{-\infty}^{\infty} \int_{-\infty}^{\infty} \int_{-\infty}^{\infty} dxdydz \mathbf{E}(x,y,z) \cdot \mathbf{J}^*(x,y,z) \tag{19}$$

$$= \frac{c}{2} \int_{-\infty}^{\infty} \int_{-\infty}^{\infty} dxdy N\pi k w_0^2 \exp(-kb) \int_{-\infty}^{\infty} \int_{-\infty}^{\infty} dp_x dp_y$$

$$\times \left(1 - \frac{4\pi^2 p_x^2}{k^2}\right) \exp\left[-i2\pi(p_x x + p_y y)\right] \zeta^{-1} \exp(\zeta b)$$

$$\times 2N\pi w_0^2 \exp(-kb) \int_{-\infty}^{\infty} \int_{-\infty}^{\infty} d\bar{p}_x d\bar{p}_y$$

$$\times \exp[i2\pi(\bar{p}_x x + \bar{p}_y y)] \exp(\bar{\zeta}^* b) \tag{20}$$

The same procedure that was used in obtaining Eq. (9) is employed. Then, P_C given by Eq. (20) reduces as

$$P_C = P_{re} + iP_{im} = cN^2\pi^2 k w_0^4 \exp(-2kb) \int_{-\infty}^{\infty} \int_{-\infty}^{\infty} dp_x dp_y$$

$$\times \left(1 - \frac{4\pi^2 p_x^2}{k^2}\right) \zeta^{-1} \exp[b(\zeta + \zeta^*)] \tag{21}$$

The value of N^2 is substituted from Eq. (1.2). Then, in the limit of the paraxial approximation, the time-averaged power transported by the corresponding fundamental Gaussian beam in the $\pm z$ direction equals $P_0^\pm = 1$ W. Equation (21) is simplified in the same manner as in obtaining Eqs. (2.21), (2.22), (2.25), and (2.26) from Eq. (2.15). The real part of Eq. (21), that is, P_{re}, is the part of the complex power P_C that corresponds to real values of ζ. P_{re} is found to be equal to $2P_b^\pm$, where $P_b^\pm$ is given by Eq. (9). Hence,

$$P_{re} = 2P_b^\pm = P_b^+ + P_b^- \tag{22}$$

Therefore, the real power is equal to the total time-averaged power transported by the basic full Gaussian wave in the $+z$ and the $-z$ directions. The real power is the time-averaged power generated by the current source and is determined by volume integration over the entire source current distributions.

After the value of N^2 is substituted from Eq. (1.2) and the integration variables are changed as given by Eq. (2.16), Eq. (21) is transformed as

$$P_C = P_{re} + iP_{im} = \frac{w_0^2}{\pi} \exp(-2kb) \int_0^{\infty} dpp \int_0^{2\pi} d\phi$$

$$\times \left(1 - \frac{p^2\cos^2\phi}{k^2}\right) \xi^{-1} \exp[kb(\xi + \xi^*)] \tag{23}$$

where ξ is defined by Eq. (2.18). The value of the integral with respect to p is imaginary for $k < p < \infty$. The integration with respect to ϕ is carried out. Then, the use of Eq. (2.18) in Eq. (23) leads to the reactive power as

$$P_{im} = -2w_0^2 \exp\left(-k^2 w_0^2\right) \int_k^\infty dp\, p \left(1 - \frac{p^2}{2k^2}\right) \left(\frac{p^2}{k^2} - 1\right)^{-1/2} \qquad (24)$$

The variable of integration is changed as given by Eq. (2.34). Then, Eq. (24) simplifies as

$$P_{im} = -k^2 w_0^2 \exp(-k^2 w_0^2) \int_0^\infty d\tau (1 - \tau^2) = \infty \qquad (25)$$

The reactive power of the basic full Gaussian wave is infinite.

The result that $P_{im} = \infty$ for the basic full Gaussian wave is not unexpected. For a point electric dipole of a given current moment in the physical space, the real power is finite, but the reactive power is infinite [6]. Omission of the small finite dimensions of the source causes the reactive power to become infinite. When the finite dimensions of the electric dipole are taken into account, the reactive power is finite and increases rapidly as the size of the electric dipole is reduced. For the basic full Gaussian wave, the virtual source is a point source, but with the location coordinates in the complex space. If the point source in the complex space is generalized to become a current distribution of finite width, it is to be expected that the reactive power would become finite and would increase rapidly as the effective width of the source is reduced. The fact that the reactive power is finite for the fundamental Gaussian wave treated in Chapter 2 lends some support to the stated view. Therefore, in order to obtain finite reactive power, the basic full Gaussian wave generated by a point source in the complex space would have to be generalized to become a full Gaussian wave associated with a source having a distribution of current in the complex space.

5 General remarks

Deschamps and Felsen have given references to the related investigations of the complex-value generalization of the geometrical optics [7–9]. Kravtsov [7] has constructed complex-value solutions of the ray equations and has treated the diffraction spreading of the fundamental Gaussian beam. Arnaud [8] has carried out analytically the method of resonance excitation of degenerate optical cavities by the use of a complex ray representation of Gaussian beams. Keller and Streifer [9] have analyzed the fundamental Gaussian beam in terms of complex rays. They use complex rays for finding the field in the region $z > 0$ when the field is specified on the plane $z = 0$. The method is applied to a Gaussian field in the plane $z = 0$, and the Gaussian beam for $z > 0$ is obtained. The calculations are carried out only for the Gaussian field on $z = 0$ in the paraxial approximation. In contrast, Deschamps and Felsen give a treatment of the full-wave generalization of the fundamental Gaussian beam by the introduction of a localized point source with the location coordinates in the complex space.

The basic full Gaussian wave generated by a point source with the location coordinates in the complex space has found some applications. For the usual treatment of open optical resonators, the light inside the cavity is modeled by the fundamental Gaussian beam multiply reflected by the two end mirrors. Cullen and Yu [10] have provided an improved treatment by modeling the light inside the cavity by the linearly polarized basic full Gaussian wave multiply reflected by the two end spherical reflectors. The basic full Gaussian wave has also been used to solve the electromagnetic problems of refraction, diffraction, and scattering [1]. In the usual treatments, the incident wave is assumed to be a plane wave. Since unbounded plane waves are not physically realizable, the treatments that use the basic full Gaussian wave as the incident wave are more realistic since the total time-averaged power radiated by the basic full Gaussian wave is finite.

A drawback of the basic full Gaussian wave is that the reactive power associated with that full wave is infinite. As pointed out before, in order to obtain finite reactive power, the basic full Gaussian wave would have to be generalized to yield a full Gaussian wave associated with a source having a distribution of current in the complex space. A class of such extended full-wave generalizations of the fundamental Gaussian beam is obtained in Chapter 6.

References

1. G. A. Deschamps, "Gaussian beam as a bundle of complex rays," *Electron. Lett.* **7**, 684–685 (1971).

2. L. B. Felsen, "Complex-source-point solutions of the field equations and their relation to the propagation and scattering of Gaussian beams," *Symposia Matematica, Istituto Nazionale di Alta Matematica* (Academic, 1976), Vol. XVIII, pp. 40–56.

3. L. B. Felsen, "Evanescent waves," *J. Opt. Soc. Am.* **66**, 751–760 (1976).

4. S. R. Seshadri, "Dynamics of the linearly polarized fundamental Gaussian light wave," *J. Opt. Soc. Am. A* **24**, 482–492 (2007).

5. S. R. Seshadri, "Quality of paraxial electromagnetic beams," *Appl. Opt.* **45**, 5335–5345 (2006).

6. S. R. Seshadri, "Constituents of power of an electric dipole of finite size," *J. Opt. Soc. Am. A* **25**, 805–810 (2008).

7. Yu. A. Kravtsov, "Complex rays and complex caustics," *Radiophys. Quantum Electron.* **10**, 719–730 (1967).

8. J. A. Arnaud, "Degenerate optical cavities. II. Effects of misalignments," *Appl. Opt.* **8**, 1909–1917 (1969).

9. J. B. Keller and W. Streifer, "Complex rays with an application to Gaussian beams," *J. Opt. Soc. Am.* **61**, 40–43 (1971).

10. A. L. Cullen and P. K. Yu, "Complex source-point theory of the electromagnetic open resonator," *Proc. R. Soc. London, Ser. A* **366**, 155–171 (1979).

CHAPTER 5

Complex source point theory

Deschamps [1] and Felsen [2] introduced the scalar point source with the location coordinates in the complex space for obtaining the exact full wave corresponding to the fundamental Gaussian beam. Postulation of the required source in the complex space for paraxial beams other than the fundamental Gaussian beam is difficult. For the purpose of generalizing the treatments of Deschamps and Felsen to any paraxial beam, a method is developed for deducing the required source with the location coordinates in the complex space [3]. In this chapter, for the fundamental Gaussian beam, such a method is presented, and the location and the strength of the required point source in the complex space are derived. Since the paraxial wave equation is only an approximation for the full Helmholtz wave equation, the same source in the complex space derived for the paraxial beam, except for an excitation coefficient as a factor, is used for the full Helmholtz equation to derive the full-wave generalization of the fundamental Gaussian beam for the two physical spaces $z > 0$ and $z < 0$. First the asymptotic $(|z| \to \infty)$ value of the full wave is determined. The asymptotic value of the full wave is analytically continued from $|z|$ to $(|z| - ib)$, and the full wave for $z > 0$ and $z < 0$ is obtained. The details of the analytic continuation are presented. The full wave is also valid for $z = 0$ provided the domain of ρ is extended to complex values and ρ is analytically continued by way of the lower half of the complex ρ plane around the branch point that exists on the real axis of the complex ρ plane. The principle of limiting absorption is introduced and used to confirm the restriction imposed on ρ for the validity of the basic full Gaussian wave for $z = 0$. The requirement in the paraxial approximation that the generated basic full Gaussian wave should reproduce exactly the initially chosen paraxial beam yields the excitation coefficient for the full wave.

1 Derivation of complex space source

The outward propagations in the $+z$ direction in $z > 0$ and in the $-z$ direction in $z < 0$ are considered. The secondary source is an infinitesimally thin sheet of electric current located on the plane $z = 0$ that forms the boundary between the two physical spaces $z > 0$ and $z < 0$. The source electric current density is expressed as

$$\mathbf{J}_e(x, y, z) = \mathbf{J}_{es}(x, y)\delta(z) \tag{1}$$

where $\mathbf{J}_{es}(x, y)$ is the strength of the source. The surface electric current density is in the x direction and the x component of the magnetic vector potential is excited.

The electromagnetic fields are found from the vector potential. The paraxial approximation of the x component of the magnetic vector potential on the input plane that generates the fundamental Gaussian light beam as given by Eq. (1.1) is reproduced as

$$A_{x0}^{\pm}(x, y, 0) = \frac{N}{ik} \exp\left(-\frac{x^2 + y^2}{w_0^2}\right) \tag{2}$$

The normalization constant N is stated in Eq. (1.2). Then the time-averaged power $P_0^{\pm}$ transported by the paraxial beam in the $\pm z$ direction is found as $P_0^{\pm} = 1$ W. As shown in Section 1.1, the integral representation of the slowly varying amplitude of the vector potential for $|z| > 0$ is given by

$$a_{x0}^{\pm}(x, y, z) = \frac{N}{ik} \pi w_0^2 \int_{-\infty}^{\infty} \int_{-\infty}^{\infty} dp_x dp_y \exp[-i2\pi(p_x x + p_y y)]$$
$$\times \exp\left[-\frac{\pi^2 w_0^2 (p_x^2 + p_y^2)}{q_{\pm}^2}\right] \tag{3}$$

where

$$q_{\pm} = \left(1 + i\frac{|z|}{b}\right)^{-1/2} \tag{4}$$

For the position coordinates in the two physical spaces $z > 0$ and $z < 0$, $1/q_{\pm}^2 \neq 0$, and the integrations in Eq. (3) are carried out to yield a closed-form expression for $a_{x0}^{\pm}(x, y, z)$. The rapidly varying phase factor, as given by Eq. (1.4), is included and the conventional expression [see Eq. (1.11)] for the fundamental Gaussian beam is determined.

The domain of z is extended to include complex values. The value of $a_{x0}^{\pm}(x, y, z)$ is sought for $|z| - ib = 0$. This choice for the location in the complex space is guided by the result obtained by Deschamps and Felsen for the fundamental Gaussian beam. Equations (3) and (4) yield $1/q_{\pm}^2 = 0$ and

$$C_{s0}(x, y) = \frac{N}{ik} \pi w_0^2 \delta(x)\delta(y) \quad \text{for } |z| - ib = 0 \tag{5}$$

Equation (5) represents the point source in the complex space postulated by Deschamps and Felsen but derived here from the two-dimensional Fourier integral representation of the paraxial beam. The source locations in the complex space for propagation in the $\pm z$ directions are different, but the source itself as given by Eq. (5) is the same for both directions of propagation.

From Maxwell's equations [Eqs. (D1) and (D2)] in the time domain, the instantaneous power generated by the distribution of electric current is found from Eq. (D6) as

$$P_{ge}(t) = -c \int_{V_s} dV \mathbf{E}(\mathbf{r}, t) \cdot \mathbf{J}_e(\mathbf{r}, t) \tag{6}$$

where c is the free-space electromagnetic wave velocity, $\mathbf{E}(\mathbf{r}, t)$ is the electric field, and $\mathbf{J}_e(\mathbf{r}, t)$ is the electric current density. The integration is carried out throughout the volume V_s

containing the current sources. The arguments of all the functions are $\mathbf{r}(= x, y, z)$ and t. All the quantities are real. Suppose that the real $|z|$ is changed to complex $|z| - ib$ with $b = \omega w_0^2/2c$, where $\omega/2\pi$ is the wave frequency and w_0 is the beam waist at the input plane. The parameter ω occurs in the time-harmonic Maxwell's equations. The instantaneous power generated, $P_{ge}(t)$, becomes complex and loses its physical significance. The point source with the location coordinates in the complex space first introduced by Deschamps and Felsen, and derived here in Eq. (5), is not an actual source but only a virtual source.

The point source in the complex space as stated in Eq. (5) is valid only for generating the fundamental Gaussian beam in the physical spaces $0 < z < \infty$ and $-\infty < z < 0$. The complex space point source given by Eq. (5) does not create any valid fields external to the two physical spaces.

2 Asymptotic field

Deschamps has suggested that the field in the physical space in the paraxial region is as if the point source generating the field has its location coordinates in the complex space. Felsen has indicated analytical continuation from $|z|$ to $(|z| - ib)$ as a means of interpretation of the field due to a point source with the location coordinates in the complex space. The analytical continuation would necessarily have to be made on the field at some location in the physical space. Therefore, it is required to relate the field at some point in the physical space to the point source in the complex space.

The point source situated at $|z| - ib = 0$ generates the paraxial beam for the entire physical spaces $z > 0$ and $z < 0$. If attention is restricted to the reconstruction of the field for only $|z| \to \infty$, that is, if we seek to reconstruct only the asymptotic $(|z| \to \infty)$ value of the paraxial beam, the source location moves from $|z| - ib = 0$ to $|z| = 0$. Therefore, the source found in Eq. (5), if placed at $|z| = 0$, yields only the asymptotic value of the paraxial beam. For the reconstruction of the asymptotic value of the paraxial beam, the source given by Eq. (5) is at the boundary of the physical space $|z| = 0$; therefore, the fields can be evaluated for all $|z| > 0$, but only the asymptotic $(|z| \to \infty)$ value of the field is valid. The operation $|z| \to \infty$ causes the source location to change from $|z| - ib = 0$ to $|z| = 0$; therefore, the effect of this operation is reversed by the analytical continuation from $|z|$ to $(|z| - ib)$ of the asymptotic value of the field, and the field for all $|z| > 0$ is recovered.

The analytical continuation is made only on the total field. For the paraxial beam, the complex point source is only for the reconstruction of the slowly varying amplitude of the beam, and the rapidly varying phase factor $\exp(\pm ikz)$ has to be included. Since the paraxial wave equation is only an approximation for the exact Helmholtz wave equation, the same complex point source as in Eq. (5), except for an excitation coefficient S_{ex} as a factor, should yield the full wave that is a solution of the Helmholtz wave equation. Hence, the complex point source for obtaining the full-wave generalization of the fundamental Gaussian beam is found from Eq. (5) as

$$C_s(x, y) = \frac{N}{ik} \pi w_0^2 S_{ex} \delta(x) \delta(y) \quad \text{for } |z| - ib = 0 \qquad (7)$$

The requirement in the paraxial approximation that the generated full wave should reproduce exactly the initially chosen paraxial beam leads to the excitation coefficient S_{ex} for the full wave.

The source that yields the asymptotic ($|z| \to \infty$) value of the full-wave solution of the Helmholtz wave equation is found from Eq. (7) as

$$C_{s\infty}(x, y, z) = \frac{N}{ik} \pi w_0^2 S_{ex} \delta(x) \delta(y) \delta(z) \tag{8}$$

The asymptotic value of the field, $A_x^{\pm}(x, y, |z| \to \infty)$, generated by the source given by Eq. (8) is obtained from Eqs. (A1), (A15), and (A16) as

$$A_x^{\pm}(x, y, |z| \to \infty) = \frac{N}{ik} \pi w_0^2 S_{ex} \frac{\exp\left[ik\left(x^2 + y^2 + z^2\right)^{1/2}\right]}{4\pi \left(x^2 + y^2 + z^2\right)^{1/2}} \tag{9}$$

When the asymptotic value of the full wave as given by Eq. (9) is analytically continued from $|z|$ to $(|z| - ib)$, the full-wave generalization of the fundamental Gaussian beam is determined as

$$A_x^{\pm}(x, y, z) = \frac{N}{ik} \pi w_0^2 S_{ex} \frac{\exp\left\{ik\left[x^2 + y^2 + (|z| - ib)^2\right]^{1/2}\right\}}{4\pi \left[x^2 + y^2 + (|z| - ib)^2\right]^{1/2}} \tag{10}$$

which is identical to that derived in Eq. (4.1). From Eqs. (4.1)–(4.4), it is seen that the paraxial approximation to $A_x^{\pm}(x, y, z)$ given by Eq. (10) reproduces the fundamental Gaussian beam as given by Eq. (4.4) if the excitation coefficient is the same as that given by Eq. (4.3), that is, if

$$S_{ex} = -2ik \exp(-kb) \tag{11}$$

Equation (11) is substituted into Eq. (10) and the vector potential that generates the basic full Gaussian wave is found as

$$A_x^{\pm}(x, y, z) = \frac{N}{ik} [-ib \exp(-kb)] \frac{\exp[iks_{\pm}(z)]}{s_{\pm}(z)} \tag{12}$$

where

$$s_{\pm}(z) = \left[x^2 + y^2 + (|z| - ib)^2\right]^{1/2} \tag{13}$$

The convention is to choose that branch of the square root of a complex number whose real part is positive.

3 Analytic continuation

The exact vector potential associated with the outgoing basic full Gaussian wave is given by Eqs. (12) and (13), where $\pm$ denotes that the propagation is in the $\pm z$ direction. For convenience, we set $\rho^2 = x^2 + y^2$. A wave propagating in the $+z$ direction from $z = -\infty$ to $z = \infty$ is an incoming wave for $z < 0$ and an outgoing wave for $z > 0$. Similarly, a wave propagating in the $-z$ direction from $z = \infty$ to $z = -\infty$ is an incoming wave for $z > 0$ and an outgoing wave for $z < 0$. The character of the wave changes at $z = 0$; therefore, the wave is expected to exhibit special features for $z = 0$. Hence the full-wave solutions for $z > 0$ and $z < 0$ are treated separately. Here, the treatment is restricted to outgoing waves. Consequently, the $+z$ and $-z$ directions of propagation correspond to the regions $z > 0$ and $z < 0$, respectively.

The time-dependent field corresponding to Eq. (12) is expressed as [4]

$$\tilde{A}_x^\pm(x, y, z; t) = \int_{\text{Cr}} d\omega \exp(-i\omega t) A_x^\pm(x, y, z; \omega) \tag{14}$$

An argument ω is added for the fields in the frequency domain. The tilde and the argument t are used to identify the fields in the time domain. Let $\omega = \omega_r + i\omega_i$, where ω_r and ω_i are the real and the imaginary parts of ω, respectively. The contour Cr in Eq. (14) is along the real ω_r axis on the complex ω plane. The outgoing wave corresponding to $\tilde{A}_x^\pm(x, y, z; t)$ is assumed to start at $t = 0$ for all x, y, and z, and it exists only for $t > 0$. Therefore, $\tilde{A}_x^\pm(x, y, z; t) = 0$ for $t < 0$. For $\omega_i \to \infty$, that is, for $-ib_i \to \infty$, Eqs. (12) and (13) simplify as

$$A_x^\pm(x, y, z; \omega) = \frac{N}{ik} \exp\left(-\frac{\rho^2}{w_0^2}\right) \exp\left[(i\omega_r - \omega_i)\frac{|z|}{c}\right] \quad \text{for } \omega_i \to \infty \tag{15}$$

Therefore, for both $z > 0$ and $z < 0$, $A_x^\pm(x, y, z; \omega)$ vanishes for $\omega_i \to \infty$. The vanishing of the argument of the square root in Eq. (13) shows that the following branch points exist on the complex ω plane:

$$\omega_0 = 2cw_0^{-2}(-i|z| \pm \rho) \tag{16}$$

For $z > 0$ and $z < 0$, the branch points lie in the lower ($\omega_i < 0$) half of the ω plane.

For $t < 0$, $\exp(-i\omega t) = \exp[(-i\omega_r + \omega_i)t]$ vanishes for $\omega_i \to \infty$. Since $\exp(-i\omega t)$ and $A_x^\pm(x, y, z; \omega)$ both vanish for $\omega_i \to \infty$, the integration contour Cr in Eq. (14) can be closed without changing the value of the integral by a semicircle Ci with its center at the origin and infinite radius in the upper ($\omega_i > 0$) half of the complex ω plane. Inside the closed contour formed by Cr and the infinite semicircle Ci, there are no poles or branch points; therefore, it follows that $\tilde{A}_x^\pm(x, y, z; t) = 0$ for $t < 0$. As a consequence, for $z > 0$ and $z < 0$, $A_x^\pm(x, y, z; \omega)$ obtained in Eqs. (12) and (13) by the analytic continuation from $|z|$ to $(|z| - ib)$ is valid. This validity occurs because $A_x^\pm(x, y, z; \omega)$ is analytic everywhere in the upper half of the ω plane, vanishes for $\omega_i \to \infty$, and, in particular, has no singularities such as poles and branch points in the upper half of the complex ω plane.

For $z = 0$, the branch points at $\pm\omega_0$, where $\omega_0 = 2cw_0^{-2}\rho$, lie on the ω_r axis. As a consequence, some special features are expected for $z = 0$. The contour Cr is indented from above ($\omega_i > 0$) these singularities on the ω_r axis (see Fig. 1). As for $z > 0$ and $z < 0$, and for $z = 0$ as well, since $\exp(-i\omega t)$ for $t < 0$ and $A_x^\pm(x, y, z; \omega)$, both vanish for $\omega_i \to \infty$, and the integration contour Cr in Eq. (14) can be closed without changing the value of the integral by the semicircle Ci. Inside the closed contour formed by the indented Cr and Ci, since there are no singularities, $\tilde{A}_x^\pm(x, y, z; t) = 0$ for $t < 0$. Hence, for $z = 0$ as well, $A_x^\pm(x, y, z; \omega)$ obtained in Eqs. (12) and (13) by the analytic continuation from $|z|$ to $(|z| - ib)$ is valid. The requirements for validity are the same as for $z > 0$ and $z < 0$.

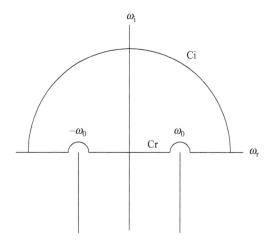

Fig. 1.   Analytical continuation around the branch points at $-\omega_0$ and ω_0 in the complex ω plane. Indented contour Cr and the semicircle Ci form a closed contour. There are no singularities inside the closed contour.

The indentation of the contour Cr from above ($\omega_i > 0$) the singularities on the ω_r axis imposes restrictions on the validity of $A_x^{\pm}(x, y, z; \omega)$ for $z = 0$. Indenting the contour Cr from above the singularities at $\omega = \pm\omega_0$, where $\omega_0 = 2cw_0^{-2}\rho$, is equivalent to requiring ω to have a small positive imaginary part near $\omega_0 = 2cw_0^{-2}\rho$ or $\rho = \omega_0 w_0^2/2c = b_0$. This requires the domain of ρ to be extended to include complex values. Let $\rho = \rho_r + i\rho_i$, where ρ_r and ρ_i are the real and the imaginary values of ρ. In the complex ρ plane, the branch point occurs at $\rho = b$. Since, near the singularity, $\omega_i > 0$, the branch point occurs in the region where $\rho_i > 0$. For obtaining physical quantities, integration with respect to ρ_r is carried out along the ρ_r axis for $0 < \rho_r < \infty$. As $\omega_i \rightarrow 0$, the branch point moves in the upper half of the ρ plane toward the ρ_r axis. To keep the ρ_r integration unchanged, the branch point is not allowed to cross the integration contour. For $\omega_i = 0$, the branch point reaches the ρ_r axis at $\rho_r = b_0$, and the contour of ρ_r integration is analytically continued by way of the lower half of the complex ρ plane. In other words, ρ is assumed to be complex with a small negative imaginary part, when ρ_r is near the branch point b_0.

4 Limiting absorption

The rectangular coordinate system (x_r, y_r, z) is used, as well as the cylindrical coordinate system (ρ_r, ϕ, z). The coordinates x_r, y_r, ρ_r, and ϕ are all real. The domain of z_r is extended to complex values $z = z_r + iz_i$, where z_r and z_i are the real and imaginary parts of z. The ranges of (x_r, y_r), ρ_r, and ϕ are given by $-\infty < (x_r, y_r) < \infty$, $0 < \rho_r < \infty$, and $0 < \phi < 2\pi$, respectively. A virtual point source of electric current density of strength S is oriented in the x_r direction and is situated at $\rho_r = 0$ and $z = ib_r$ where the real basic Rayleigh distance is given by

$$b_r = \frac{1}{2}k_r w_0^2 \tag{17}$$

The wavenumber k_r is real and is defined by $k_r = \omega(\mu\varepsilon)^{1/2}$, where μ and ε are the permeability and the permittivity of the medium, respectively. The beam waist in the paraxial approximation at the input plane $z_i = 0$ and $z_r = 0$ is given by w_0. The physical spaces are defined by $(z_i = 0, z_r > 0)$ and $(z_i = 0, z_r < 0)$. In the physical spaces, the vector potential that generates the outgoing basic full Gaussian wave is reproduced from Eqs. (12) and (13) as

$$A_x^{\pm}(x_r, y_r, z_r; \omega) = S \frac{\exp[ik_r s_{\pm}(\rho_r, z_r)]}{4\pi s_{\pm}(\rho_r, z_r)} \tag{18}$$

where the complex distance is given by

$$s_{\pm}(\rho_r, z_r) = \left[\rho_r^2 + (|z_r| - ib_r)^2\right]^{1/2} \tag{19}$$

and $S = (N/ik)[-4\pi ib \exp(-kb)]$. The restriction on the validity of $A_x^{\pm}(x_r, y_r, z_r; \omega)$ for $z_r = 0$ can also be derived from the principle of limiting absorption as follows [5].

The branch of the square root, which is a two-valued function, has to be chosen properly. When the two branches coalesce, it is not possible to separate out the proper branch. Hence, the zero of the argument of the square root is a singular point known as the branch point. Therefore, the selection of the proper branch of $s_{\pm}(\rho_r, z_r)$ includes the avoidance of the singular point.

The proper branch of $s_{\pm}(\rho_r, 0)$ is desired. The branch point is real and is given by

$$\rho_r = b_r \quad \text{for } z_r = 0 \tag{20}$$

For $\rho_r > b_r$, $s_{\pm}(\rho_r, 0)$ is real. In accordance with our convention of choosing the branch with the positive real part, the proper branch of $s_{\pm}(\rho_r, 0)$ for $\rho_r > b_r$ is found from Eq. (19) as

$$s_{\pm}(\rho_r, z_r) = (\rho_r^2 - b_r^2)^{1/2} \quad \text{for } \rho_r > b_r \tag{21}$$

For $\rho_r < b_r$, $s_{\pm}(\rho_r, 0)$ is imaginary, and our convention of choosing the proper branch of the square root is not sufficient to select the proper branch when $s_{\pm}(\rho_r, 0)$ is imaginary. For the selection of the proper branch of the square root when $\rho_r < b_r$, a small loss is included in the medium. Then ε changes to $\varepsilon(1 + it_\ell)$, where the positive real loss tangent is given by $t_\ell = \sigma/\omega\varepsilon$; the conductivity of the medium is denoted by σ. The wavenumber and the Rayleigh distance become complex, as given by

$$k = k_r + ik_i = k_r(1 + it_\ell)^{1/2} = k_r\left(1 + i\frac{1}{2}t_\ell\right) \tag{22}$$

and

$$b = b_r + ib_i = b_r\left(1 + i\frac{1}{2}t_\ell\right) \tag{23}$$

When there are losses, b_r changes to complex b. The domain of ρ_r is extended to complex $\rho = \rho_r + i\rho_i$. In the presence of losses, the branch point becomes complex, as given by

$$\rho = \rho_r + i\rho_i = b_r\left(1 + i\frac{1}{2}t_\ell\right) \quad \text{for } z_r = 0 \tag{24}$$

The real part of the branch point is positive, and the imaginary part is also positive. As explained previously, in the limit of losses reducing to zero, the branch point reaches the real ρ_r axis at $\rho_r = b_r$, and the contour of ρ_r integration is analytically continued by way of the lower half of the complex ρ plane. The contour is along the real ρ_r axis from the origin $O(\rho_r = 0)$ to point A $(\rho_r = b_r - \Delta)$ and from point B $(\rho_r = b_r + \Delta)$ to infinity $(\rho_r = \infty)$(Fig. 2). From A to B, the contour is indented along the semicircular path ACB with its center at $\rho_r = b_r$ and a small radius Δ. Along the semicircular path, $\rho_i < 0$. The complex distance at B is recast as

$$s_\pm\left(\rho_r, 0\right) = \left(\rho_r^2 - b_r^2\right)^{1/2} = \left[(2b_r + \Delta)\Delta\right]^{1/2} \tag{25}$$

As the observation point moves from B along the semicircular path BCA in the clockwise direction through an angle π to A, the complex distance at A becomes

$$\begin{aligned}
s_\pm(\rho_r, 0) &= \left\{[2b_r + \Delta\exp(-i\pi)]\Delta\exp(-i\pi)\right\}^{1/2} \\
&= \left\{[2b_r - (b_r - \rho_r)](b_r - \rho_r)\right\}^{1/2}\exp(-i\pi/2) \\
&= -i\left(b_r^2 - \rho_r^2\right)^{1/2} \quad \text{for } \rho_r < b_r
\end{aligned} \tag{26}$$

Thus, the proper branch of $s_\pm(\rho_r, 0)$ is given by Eq. (21) for $\rho_r > b_r$ and by Eq. (26) for $\rho_r < b_r$. The small distance Δ is made arbitrarily small but is not reduced to zero. Therefore, the complex distance $s_\pm(\rho_r, 0)$ never becomes zero.

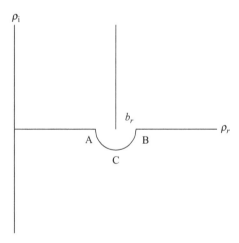

Fig. 2.   Analytical continuation around the branch point $(\rho_r = b_r)$ in the complex ρ plane.

If the observation point moves from B along a semicircular path in the counter-clockwise direction through an angle π to A, the complex distance at A becomes $s_{\pm}(\rho_r, 0) = i(b_r^2 - \rho_r^2)^{1/2}$, which is the second (improper) branch of $s_{\pm}(\rho_r, 0)$. The occurrence of the improper branch is eliminated by introducing a branch cut extending from $\rho = b_r + i0$ to $\rho = b_r + i\infty$ parallel to the ρ_i axis on the upper half of the complex ρ plane. Another plane, named the improper sheet, which is similar to the proper sheet, is introduced. The two sheets are joined along the branch cut in such a way that the observation point, on crossing the branch cut, moves from one sheet to the other. If the observation point moves from B in the counterclockwise direction, on crossing the branch cut, the observation point enters the improper sheet; after a circulation of angle 2π in the counterclockwise direction on the improper sheet, the observation point crosses the branch cut, reenters the proper sheet, and reaches A after a total rotation of angle 3π in the counterclockwise direction. Then the complex distance at A becomes the same as that given by Eq. (26). Consequently, on the proper sheet with the branch cut, only the proper branch of $s_{\pm}(\rho_r, 0)$, as defined by Eq. (26), is obtained independent of the direction of the path, clockwise or counterclockwise, from B to A.

Summing up, for $z = 0$ as well, $A_x^{\pm}(x, y, z; \omega)$ obtained in Eqs. (12) and (13) by the analytic continuation from $|z|$ to $(|z| - ib)$ is valid. The restriction is that the domain of ρ should be extended to complex values, and ρ should have a small negative imaginary part, when ρ_r is near the branch point b_r. This restriction is realized by the introduction of a branch cut from $\rho = b_r + i0$ to $\rho = b_r + i\infty$ parallel to the ρ_i axis.

References

1. G. A. Deschamps, "Gaussian beam as a bundle of complex rays," *Electron. Lett.* **7**, 684–685 (1971).

2. L. B. Felsen, "Evanescent waves," *J. Opt. Soc. Am.* **66**, 751–760 (1976).

3. S. R. Seshadri, "Linearly polarized anisotropic Gaussian light wave," *J. Opt. Soc. Am. A* **26**, 1582–1587 (2009).

4. S. R. Seshadri, "Independent waves in complex source point theory," *Opt. Lett.* **32**, 3218–3220 (2007).

5. S. R. Seshadri, "High-aperture beams: comment," *J. Opt. Soc. Am. A* **23**, 3238–3241 (2006).

CHAPTER 6

Extended full Gaussian wave

The fundamental Gaussian light beam has no reactive power [1]. For the basic full Gaussian wave, the reactive power is infinite. In order to obtain finite reactive power, a full Gaussian wave generated by a distribution of electric current rather than a point source in the complex space is necessary. A class of such extended full-wave generalizations of the fundamental Gaussian beam is obtained [2]. Various members of this class of extended full Gaussian waves are identified by a length parameter b_t that is in the range $0 \leq b_t/b \leq 1$, where b is the Rayleigh distance. The limiting case of $b_t/b = 1$ is the basic full Gaussian wave. For the limiting case of $b_t/b = 0$, the source electric current density is the same for the fundamental Gaussian beam and for the extended full Gaussian wave.

The exact vector potential that generates the extended full Gaussian wave is deduced. The relevant electromagnetic field components are derived, the radiation intensity distribution is found, and the time-averaged power transported by the extended full Gaussian wave in the $+z$ and $-z$ directions is evaluated. The dependence of the power on b_t/b and kw_0 is analyzed. The characteristics of the radiation intensity distribution, in particular its dependence on b_t/b, are discussed. The surface electric current density on the secondary source plane $z = 0$ is deduced. For all b_t/b in the range $0 \leq b_t/b \leq 1$, in the paraxial approximation, the surface electric current density reduces to the source electric current density that generates the same fundamental Gaussian beam. The complex power is evaluated and the reactive power is determined. The dependence of the reactive power on b_t/b and kw_0 is investigated.

1 Current source of finite extent in complex space

The method developed for the derivation of the full-wave generalization of the fundamental Gaussian beam permits the determination of a class of full Gaussian waves, all of which reduce to the same fundamental Gaussian beam in the paraxial approximation. The search for the required complex space source is carried out at $|z| - ib_t = 0$, where b_t is real. Then, $1/q_{\pm}^2 = 1 - b_t/b$. It is assumed that $0 \leq b_t/b \leq 1$. Then, Eq. (5.3) yields the source in the complex space as

$$C_{s0}(x,y) = \frac{N}{ik} \frac{1}{(1 - b_t/b)} \exp\left[-\frac{x^2 + y^2}{w_0^2(1 - b_t/b)}\right] \quad \text{for } |z| - ib_t = 0 \qquad (1)$$

Writing Eq. (1) as

$$C_{s0}(x, y) = \frac{N}{ik} \pi w_0^2 C_{sx}(x) C_{sy}(y) \quad \text{for } |z| - ib_t = 0 \tag{2}$$

where

$$C_{sx}(x) = \frac{1}{\sqrt{\pi} w_0 (1 - b_t/b)^{1/2}} \exp\left[-\frac{x^2}{w_0^2(1 - b_t/b)}\right] \tag{3}$$

we note that $\int_{-\infty}^{\infty} dx C_{sx}(x) = 1$. The mean squared width of $C_{sx}(x)$ is found as

$$\int_{-\infty}^{\infty} dx x^2 C_{sx}(x) = \frac{1}{2} w_0^2 \left(1 - \frac{b_t}{b}\right) \tag{4}$$

The mean squared width of the source distribution in the x direction is largest for $b_t = 0$, becomes smaller as b_t is increased, and becomes smallest for $b_t = b$. Similar characteristics are obtained for the source distribution in the y direction. Therefore, the mean width of the source distribution decreases as b_t increases from 0 to b. The rate of spreading of the beam on propagation increases as the width of the source distribution is decreased. Therefore, on propagation, wider beams are to be expected for $b_t = b$ than for $b_t = 0$. Research on the full-wave generalization of the fundamental Gaussian beam was initiated in part with a view to explaining satisfactorily the large beam widths observed in some laser outputs, the paraxial approximation not being sufficient for this purpose. Hence, the reasons for the introduction of the basic full Gaussian wave are valid. The reactive power of the basic full Gaussian wave is infinite. In order to obtain finite reactive power, it is necessary that the source distribution in the complex space be of finite extent. Consequently, the treatment of the full Gaussian waves for $b_t < b$ leading to a source current distribution in the complex space of finite extent is warranted.

The source given by Eq. (1) located at $|z| - ib_t = 0$ generates the paraxial beam for all $|z| > 0$. For $|z| \to \infty$, the source location moves from $|z| - ib_t = 0$ to $|z| = 0$. Therefore, the source given by Eq. (1) is moved to $|z| = 0$, that is, to the boundary of the physical space $|z| > 0$, for the determination of the asymptotic ($|z| \to \infty$) value of the paraxial beam. Since, for $|z| \to \infty$, the complex space source is moved from $|z| - ib_t = 0$ to $|z| = 0$, the effect of moving the source back from $|z| = 0$ to $|z| - ib_t = 0$, that is, to the original location, is recovered by the analytical continuation of the asymptotic value of the paraxial beam from $|z|$ to $|z| - ib_t$. For the paraxial beam, the asymptotic value pertains only to the slowly varying amplitude. But the analytical continuation has to be carried out on the total field, which includes also the rapidly varying plane-wave phase: $\exp(\pm ikz)$. Since the paraxial wave equation is only an approximation for the full Helmholtz wave equation, the complex space source deduced from the paraxial beam, except for an excitation coefficient S_{ex}, is used for the Helmholtz equation to obtain the asymptotic value of the full Gaussian wave. Hence, the complex space source for obtaining the full-wave generalization of the fundamental Gaussian beam is found from Eq. (1) as

$$C_s(x, y) = \frac{N}{ik} \frac{S_{ex}}{(1 - b_t/b)} \exp\left[-\frac{x^2 + y^2}{w_0^2(1 - b_t/b)}\right] \quad \text{for } |z| - ib_t = 0 \tag{5}$$

As before, the requirement in the paraxial approximation that the excited full wave should reduce exactly to the initially chosen paraxial beam allows the evaluation of the excitation coefficient for the full wave.

The source that yields the asymptotic value of the full-wave solution of the Helmholtz wave equation is obtained from Eq. (5) as

$$C_{s\infty}(x,y,z) = \frac{N}{ik}\frac{S_{ex}}{(1 - b_t/b)} \exp\left[-\frac{x^2 + y^2}{w_0^2(1 - b_t/b)}\right]\delta(z) \tag{6}$$

Let $G(x,y,z)$ be the solution of the Helmholtz equation for the source term given by Eq. (6). Then, $G(x,y,z)$ satisfies the following inhomogeneous Helmholtz equation:

$$\left(\frac{\partial^2}{\partial x^2} + \frac{\partial^2}{\partial y^2} + \frac{\partial^2}{\partial z^2} + k^2\right)G(x,y,z) = -\frac{N}{ik}\frac{S_{ex}}{(1 - b_t/b)} \exp\left[-\frac{x^2 + y^2}{w_0^2(1 - b_t/b)}\right]\delta(z) \tag{7}$$

$G(x,y,z)$ is replaced by its two-dimensional Fourier integral representation [see Eq. (A17)]. The right-hand side of Eq. (7) divided by $-S_{ex}\delta(z)$ is equal to $C_{s0}(x,y)$. $C_{s0}(x,y)$ is replaced by its two-dimensional inverse Fourier transform. The Fourier transform of $G(x,y,z)$ is denoted by $\overline{G}(p_x, p_y, z)$. From Eq. (7), $\overline{G}(p_x, p_y, z)$ is found to satisfy the following differential equation:

$$\left(\frac{\partial^2}{\partial z^2} + \zeta^2\right)\overline{G}(p_x, p_y, z) = -S_{ex}\frac{N}{ik}\pi w_0^2 \exp\left[-\pi^2 w_0^2\left(p_x^2 + p_y^2\right)\left(1 - \frac{b_t}{b}\right)\right]\delta(z) \tag{8}$$

where

$$\zeta = \left[k^2 - 4\pi^2\left(p_x^2 + p_y^2\right)\right]^{1/2} \tag{9}$$

See Eq. (2.3). With the help of Eqs. (A19)–(A26), the solution of Eq. (8) is expressed as

$$\overline{G}(p_x, p_y, z) = S_{ex}\frac{N}{2k}\pi w_0^2 \exp\left[-\pi^2 w_0^2\left(p_x^2 + p_y^2\right)\left(1 - \frac{b_t}{b}\right)\right]\zeta^{-1}\exp(i\zeta|z|) \tag{10}$$

$\overline{G}(p_x, p_y, z)$ is analytically continued from $|z|$ to $|z| - ib_t$ to find the Fourier transform of the vector potential as

$$\overline{A}_x^{\pm}(p_x, p_y, z) = S_{ex}\frac{N}{2k}\pi w_0^2 \exp\left[-\pi^2 w_0^2\left(p_x^2 + p_y^2\right)\left(1 - \frac{b_t}{b}\right)\right]$$
$$\times \zeta^{-1}\exp[i\zeta(|z| - ib_t)] \tag{11}$$

The paraxial approximation corresponds to $4\pi^2(p_x^2 + p_y^2) \ll k^2$. Then, as given by Eq. (C17), ζ can be expanded as

$$\zeta = k - \frac{\pi^2 w_0^2(p_x^2 + p_y^2)}{b} \tag{12}$$

In Eq. (11), if ζ in the amplitude is replaced by the first term in Eq. (12) and ζ in the phase is replaced by the first two terms in Eq. (12), the paraxial approximation of Eq. (11) is found as

$$\overline{A}_x^{\pm}(p_x, p_y, z) = \exp(\pm ikz) \frac{S_{ex} \pi w_0^2 \exp(kb_t)}{2k} \frac{N}{k}$$

$$\times \exp\left[-\frac{\pi^2 w_0^2 (p_x^2 + p_y^2)}{q_{\pm}^2}\right] \tag{13}$$

where $q_{\pm}$ is defined by Eq. (5.4). The excitation coefficient is chosen as

$$S_{ex} = -2ik \exp(-kb_t) \tag{14}$$

Then, Eq. (13) simplifies as

$$\overline{A}_x^{\pm}(p_x, p_y, z) = \exp(\pm ikz) \frac{N}{ik} \pi w_0^2 \exp\left[-\frac{\pi^2 w_0^2 (p_x^2 + p_y^2)}{q_{\pm}^2}\right] \tag{15}$$

The inverse Fourier transform of Eq. (15) is found as

$$A_x^{\pm}(x, y, z) = \exp(\pm ikz) \frac{N}{ik} q_{\pm}^2 \exp\left[-\frac{q_{\pm}^2 (x^2 + y^2)}{w_0^2}\right] \tag{16}$$

which is identical to that given by Eq. (4.4). If the excitation coefficient is given by Eq. (14), $\overline{A}_x^{\pm}(p_x, p_y, z)$ in Eq. (11) reproduces correctly the fundamental Gaussian beam as given by Eq. (16). $\overline{A}_x^{\pm}(p_x, p_y, z)$ is simplified by substituting Eq. (14) into Eq. (11). The inverse Fourier transform of $\overline{A}_x^{\pm}(p_x, p_y, z)$ yields the vector potential governing the full-wave generalization of the fundamental Gaussian beam as

$$A_x^{\pm}(x, y, z) = \frac{N}{i} \pi w_0^2 \exp(-kb_t) \int_{-\infty}^{\infty} \int_{-\infty}^{\infty} dp_x dp_y \exp[-i2\pi(p_x x + p_y y)]$$

$$\times \exp\left[-\pi^2 w_0^2 \left(p_x^2 + p_y^2\right)\left(1 - \frac{b_t}{b}\right)\right] \zeta^{-1} \exp[i\zeta(|z| - ib_t)] \tag{17}$$

The full wave generated by the vector potential in Eq. (17) is designated the extended full Gaussian wave. Equation (17) is identical to Eq. (4.6) for $b_t = b$ and to Eq. (2.5) for $b_t = 0$. The fundamental Gaussian wave ($b_t = 0$) and the basic full Gaussian wave ($b_t = b$) form the limiting cases of a class of extended full Gaussian waves ($0 \leq b_t \leq b$), all of which reduce to the same fundamental Gaussian beam in the paraxial approximation.

2 Time-averaged power

The electromagnetic field components $E_x^{\pm}(x, y, z)$ and $H_y^{\pm}(x, y, z)$, which are required to obtain the Poynting vector and the complex power, are found from Eq. (17), (D30), and (D33) as

$$E_x^{\pm}(x, y, z) = N\pi k w_0^2 \exp(-kb_t) \int_{-\infty}^{\infty} \int_{-\infty}^{\infty} dp_x dp_y$$

$$\times \left(1 - \frac{4\pi^2 p_x^2}{k^2}\right) \exp[-i2\pi(p_x x + p_y y)]$$

$$\times \exp\left[-\pi^2 w_0^2 \left(p_x^2 + p_y^2\right)\left(1 - \frac{b_t}{b}\right)\right]$$

$$\times \zeta^{-1} \exp[i\zeta(|z| - ib_t)] \tag{18}$$

and

$$H_y^{\pm}(x, y, z) = \pm N\pi w_0^2 \exp(-kb_t) \int_{-\infty}^{\infty} \int_{-\infty}^{\infty} d\overline{p}_x d\overline{p}_y$$

$$\times \exp[-i2\pi(\overline{p}_x x + \overline{p}_y y)]$$

$$\times \exp\left[-\pi^2 w_0^2 \left(\overline{p}_x^2 + \overline{p}_y^2\right)\left(1 - \frac{b_t}{b}\right)\right]$$

$$\times \exp[i\overline{\zeta}(|z| - ib_t)] \tag{19}$$

The time-averaged power flow per unit area in the $\pm z$ direction, $[\pm S_z^{\pm}(x, y, z)]$, is obtained by substituting Eqs. (18) and (19) into Eq. (1.13). The time-averaged power $P_e^{\pm}$ transported by the extended full Gaussian wave in the $\pm z$ direction is determined by integrating $[\pm S_z^{\pm}(x, y, z)]$ with respect to x and y across the entire transverse plane. The expression for $P_e^{\pm}$ is transformed by the use of a procedure similar to that employed in obtaining Eq. (2.15), with the result that

$$P_e^{\pm} = \frac{c}{2} N^2 \pi^2 k w_0^4 \exp(-2kb_t) \mathrm{Re} \int_{-\infty}^{\infty} \int_{-\infty}^{\infty} dp_x dp_y$$

$$\times \left(1 - \frac{4\pi^2 p_x^2}{k^2}\right) \exp\left[-2\pi^2 w_0^2 \left(p_x^2 + p_y^2\right)\left(1 - \frac{b_t}{b}\right)\right]$$

$$\times \zeta^{-1} \exp[i|z|(\zeta - \zeta^*)] \exp[b_t(\zeta + \zeta^*)] \tag{20}$$

The value of N^2 is substituted from Eq. (1.2). Then, $P_e^{\pm}$, in the limit of the paraxial approximation, equals $P_0^{\pm} = 1$ W. Equation (20) is manipulated in the same manner as in obtaining Eqs. (2.21), (2.22), (2.25), and (2.26) from Eq. (2.15) to yield [3]

$$P_e^{\pm} = \int_0^{\pi/2} d\theta^{\pm} \sin\theta^{\pm} \int_0^{2\pi} d\phi \Phi_e(\theta^{\pm}, \phi) \tag{21}$$

where

$$\Phi_e(\theta^{\pm}, \phi) = \frac{(1 - \sin^2\theta^{\pm} \cos^2\phi)}{2\pi f_0^2} \exp\left[-\frac{1}{2} k^2 w_0^2 \left(1 - \frac{b_t}{b}\right) \sin^2\theta^{\pm}\right]$$

$$\times \exp\left[-k^2 w_0^2 \frac{b_t}{b}(1 - \cos\theta^{\pm})\right] \tag{22}$$

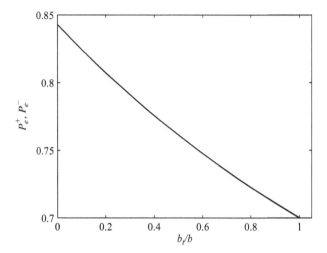

Fig. 1.   Time-averaged power $P_e^{\pm}$ transported by the extended full Gaussian wave in the $\pm z$ direction as a function of b_t/b for $0 \leq b_t/b \leq 1$ and $kw_0 = 1.563$. Time-averaged power in the corresponding paraxial beam is given by $P_0^{\pm} = 1$ W.

and as defined by Eq. (2.24),

$$\theta^- = \pi - \theta^+ \tag{23}$$

It follows from Eqs. (21) and (22) that $P_e^+ = P_e^-$.

The extended full Gaussian wave is seen from Eq. (22) to be characterized by two parameters, $kw_0 (= 1/f_0)$ and b_t/b. Equation (22) is substituted into Eq. (21), the ϕ integration is carried out analytically, and the $\theta^{\pm}$ integration is performed numerically to obtain $P_e^{\pm}$ as a function of b_t/b for a fixed kw_0 and as a function of kw_0 for a fixed b_t/b. In Fig. 1, for $kw_0 = 1.563$, $P_e^{\pm}$ is shown as a function of b_t/b for $0 \leq b_t/b \leq 1$. The limiting cases of $b_t/b = 1$ and 0 correspond to the basic full Gaussian wave and the fundamental Gaussian wave, respectively. $P_e^{\pm}$ increases monotonically as b_t/b is decreased from 1 to 0. Since for the corresponding paraxial beam, namely the fundamental Gaussian beam, the time-averaged power transported in the $\pm z$ direction is given by $P_0^{\pm} = 1$ W, Fig. 1 shows that the power ratio $P_0^{\pm}/P_e^{\pm}$ approaches 1 as b_t/b is reduced from 1 to 0. Therefore, the quality of the paraxial approximation to the full wave improves as b_t/b is decreased from 1 to 0 [4]. In particular, the fundamental Gaussian beam approximates the fundamental Gaussian wave better than it does the basic full Gaussian wave. In Fig. 2, for $b_t/b = 0.5$, $P_e^{\pm}$ is shown as a function of kw_0 for $1 < kw_0 < 10$. $P_e^{\pm}$ increases monotonically and approaches the limiting value of $P_0^{\pm} = 1$ W corresponding to the fundamental Gaussian beam as kw_0 is increased. Since $P_e^{\pm}/P_0^{\pm}$ approaches 1 as kw_0 is increased, the quality of the paraxial beam approximation to the full wave improves as kw_0 is increased [4].

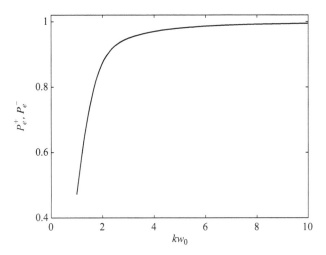

Fig. 2. Time-averaged power $P_e^\pm$ transported by the extended full Gaussian wave in the $\pm z$ direction as a function of kw_0 for $1 < kw_0 < 10$ and $b_t/b = 0.5$. Time-averaged power in the corresponding paraxial beam is $P_0^\pm = 1$ W.

3 Radiation intensity

It follows from Eqs. (21) and (22) that $\Phi_e(\theta^\pm, \phi)$ is the time-averaged power flow per unit solid angle in the direction specified by $(\theta^\pm, \phi)$; hence, $\Phi_e(\theta^+, \phi)$ and $\Phi_e(\theta^-, \phi)$ are the radiation intensities for $z > 0$ and $z < 0$, respectively. As mentioned previously, for $z > 0$, θ^+ is defined with respect to the $+z$ axis, and for $z < 0$, θ^- is defined with respect to the $-z$ axis. In view of Eq. (23), $\Phi_e(\theta^-, \phi)$ for $z < 0$ is the mirror image of $\Phi_e(\theta^+, \phi)$ for $z > 0$ about the $z = 0$ plane. Therefore, only the radiation intensity distribution for $z > 0$ is discussed.

The radiation intensity $\Phi_e(\theta^+, \phi)$ is a function of ϕ but has reflection symmetries about the $\phi = (0°, 180°)$ and $\phi = (90°, 270°)$ planes; therefore, the variation of $\Phi_e(\theta^+, \phi)$ for one quadrant only, namely for $0° < \phi < 90°$, is treated. In every azimuthal plane $\phi =$ constant, $\Phi_e(\theta^+, \phi)$ has the value $1/2\pi f_0^2$ for $\theta^+ = 0°$, decreases monotonically as θ^+ increases, and reaches a minimum value for $\theta^+ = 90°$. This minimum value is found from Eqs. (21) and (22) as

$$\Phi_e(\theta^+ = 90°, \phi) = \frac{\sin^2 \phi}{2\pi f_0^2} \exp\left[-\frac{1}{2} k^2 w_0^2 \left(1 + \frac{b_t}{b}\right)\right] \qquad (24)$$

This minimum value increases from 0 for $\phi = 0°$ to a maximum value for $\phi = 90°$. The beam width of the extended full Gaussian wave has a minimum value for $\phi = 0°$, increases continuously as ϕ is increased, and reaches a maximum value for $\phi = 90°$.

In Fig. 3, the radiation intensity patterns $\Phi_e(\theta^+, \phi)$ of the extended full Gaussian wave are depicted as functions of θ^+ for $0° < \theta^+ < 90°$, $kw_0 = 1.563$, $b_t/b = 0.5$, and for $\phi = 0°$, 30°, 60°, and 90°. The radiation intensity pattern $\Phi_0(\theta^+, \phi)$ of the corresponding fundamental Gaussian beam is also shown for the purpose of comparison. $\Phi_0(\theta^+, \phi)$ is independent of ϕ. The time-averaged power $P_0^\pm$ in the fundamental Gaussian beam propagating in the $+z$ or the $-z$ direction is 1 W. $\Phi_0(\theta^+, \phi)$ is a good approximation to $\Phi_e(\theta^+, \phi)$ for all ϕ and for $0° < \theta^+ < 90°$. The quality of the agreement between $\Phi_e(\theta^+, \phi)$ and $\Phi_0(\theta^+, \phi)$ improves as ϕ is increased from $0°$ to $90°$. Consequently, the differences between $\Phi_e(\theta^+, \phi)$ and $\Phi_0(\theta^+, \phi)$ are largest for $\phi = 0°$.

A comparison of Fig. 3 with Figs. 2.2 and 4.2 shows that the general features of the radiation intensity pattern of the extended full Gaussian wave for $b_t/b = 0.5$ are the same as those for the fundamental Gaussian wave and the basic full Gaussian wave. As pointed out previously, the fundamental Gaussian wave and the basic full Gaussian wave are also special and limiting cases of the extended full Gaussian wave corresponding to $b_t/b = 0$ and $b_t/b = 1$, respectively. Therefore, the radiation intensity patterns of the extended full Gaussian wave do not change significantly as the parameter b_t/b is changed from 0 to 1. All the extended full Gaussian waves corresponding to the range of values of b_t/b from 0 to 1 reduce to the same fundamental Gaussian beam in the paraxial approximation. The differences in the radiation intensity patterns of the different members of the class of extended full Gaussian waves become noticeable as θ^+ increases toward $\theta^+ = 90°$. The contribution of the exponential terms to the radiation intensity decreases rapidly as θ^+ is increased from $0°$ to $90°$. This feature reduces the differences in the radiation intensity patterns pertaining to the different values of b_t/b for θ^+ near $90°$. With a view to bringing out these characteristics of the radiation intensity patterns for different values of b_t/b, in Fig. 4 the radiation intensity patterns $\Phi_e(\theta^+, \phi)$ of the extended full Gaussian wave are shown as functions of θ^+ for $0° < \theta^+ < 90°$, for $\phi = 0°$, $kw_0 = 1.563$, and for the two extreme values of b_t/b, namely 0 and 1. Figure 4 reveals that for the entire range of θ^+ from $0°$ to $90°$, the radiation intensity

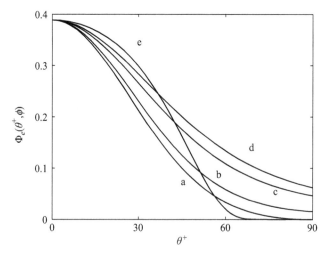

Fig. 3.   Radiation intensity pattern $\Phi_e(\theta^+, \phi)$ of the extended full Gaussian wave for curves (a) $\phi = 0°$, (b) $\phi = 30°$, (c) $\phi = 60°$, (d) $\phi = 90°$, and (e) cylindrically symmetric radiation intensity pattern $\Phi_0(\theta^+, \phi)$ of the corresponding paraxial Gaussian beam as functions of θ^+ for $0° < \theta^+ < 90°$. Other parameters are $kw_0 = 1.563$, $b_t/b = 0.5$; the power in the paraxial Gaussian beam is 1 W, and that in the extended full Gaussian wave is 0.7612 W.

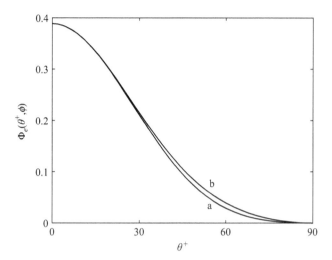

Fig. 4. Radiation intensity pattern $\Phi_e(\theta^+,\phi)$ of the extended full Gaussian wave for curves (a) $b_t/b = 1$ and (b) $b_t/b = 0$ as functions of θ^+ for $0° < \theta^+ < 90°$. Other parameters are $kw_0 = 1.563$, $\phi = 0°$; the power in the paraxial Gaussian beam is 1 W, and that in the extended full Gaussian wave is 0.7001 W for $b_t/b = 1$ and 0.8428 W for $b_t/b = 0$.

distribution $\Phi_e(\theta^+,\phi)$ of the extended full Gaussian wave is fairly insensitive to the changes in b_t/b from 0 to 1.

4 Radiative and reactive powers

$A_x^\pm(x,y,z)$ and $E_x^\pm(x,y,z)$, given by Eqs. (17) and (18), respectively, are continuous across the $z = 0$ plane. But $H_y^\pm(x,y,z)$ given by Eq. (19) has a discontinuity across the $z = 0$ plane. The discontinuity of $H_y^\pm(x,y,z)$ is equivalent to a surface electric current density on the plane $z = 0$. The surface electric current density is obtained from Eq. (19) as

$$\mathbf{J}(x,y,z) = \hat{z} \times \hat{y}[H_y^+(x,y,0) - H_y^-(x,y,0)]$$

$$= -\hat{x}2N\pi w_0^2 \exp(-kb_t)\delta(z)\int_{-\infty}^{\infty}\int_{-\infty}^{\infty} d\bar{p}_x d\bar{p}_y \exp[-i2\pi(\bar{p}_x x + \bar{p}_y y)]$$

$$\times \exp\left[-\pi^2 w_0^2\left(\bar{p}_x^2 + \bar{p}_y^2\right)\left(1 - \frac{b_t}{b}\right)\right]\exp\left[\bar{\xi}b_t\right] \tag{25}$$

The complex power is found by substituting Eqs. (18) and (25) in Eq. (D18) as

$$P_C = cN^2\pi^2 kw_0^4 \exp(-2kb_t)\int_{-\infty}^{\infty}\int_{-\infty}^{\infty} dp_x dp_y$$

$$\times\left(1 - \frac{4\pi^2 p_x^2}{k^2}\right)\exp\left[-2\pi^2 w_0^2\left(p_x^2 + p_y^2\right)\left(1 - \frac{b_t}{b}\right)\right]$$

$$\times \zeta^{-1}\exp[b_t(\zeta + \zeta^*)] \tag{26}$$

The same procedure that is used in deriving Eq. (20) is also employed in obtaining Eq. (26). The value of N^2 is substituted from Eq. (1.2). Then, in the limit of the paraxial approximation, the time-averaged power transported by the corresponding fundamental Gaussian beam in the $\pm z$ direction equals $P_0^\pm = 1$ W. Equation (26) is transformed in the same manner as in deriving Eqs. (2.21), (2.22), (2.25), and (2.26) from Eq. (2.15). The real power P_{re} of Eq. (26) is the part of the complex power P_C that corresponds to real values of ζ. P_{re} is found to be equal to $2P_e^\pm$, where $P_e^\pm$ is given by Eqs. (21)–(23). Therefore,

$$P_{re} = 2P_e^\pm = P_e^+ + P_e^- \tag{27}$$

Hence, the real power is equal to the total time-averaged power transported by the extended full Gaussian wave in the $+z$ and $-z$ directions. The real power is the time-averaged power generated by the electric current source distribution; it is determined by the integration over the entire secondary source plane $z = 0$ where the electric current exists.

The value of N^2 is substituted from Eq. (1.2) and the integration variables are changed as given by Eq. (2.16). Then, Eq. (26) is transformed as

$$P_C = P_{re} + iP_{im} = \frac{w_0^2}{\pi} \exp(-2kb_t) \int_0^\infty dp\, p \int_0^{2\pi} d\phi \left(1 - \frac{p^2 \cos^2 \phi}{k^2} \right)$$
$$\times \exp\left[-\frac{w_0^2}{2} \left(1 - \frac{b_t}{b} \right) p^2 \right] \xi^{-1} \exp[kb_t(\xi + \xi^*)] \tag{28}$$

where [see Eq. (2.18)] $\xi = (1 - p^2/k^2)^{1/2}$. For $k < p < \infty$, ξ is imaginary. The imaginary part of P_C, or the reactive power, is found from Eq. (28) as

$$P_{im} = -\frac{w_0^2}{\pi} \exp(-2kb_t) \int_k^\infty dp\, p \int_0^{2\pi} d\phi \left(1 - \frac{p^2 \cos^2 \phi}{k^2} \right)$$
$$\times \exp\left[-\frac{w_0^2}{2} \left(1 - \frac{b_t}{b} \right) p^2 \right] \xi_{\text{Im}}^{-1} \tag{29}$$

where

$$\xi_{\text{Im}} = (p^2/k^2 - 1)^{1/2} \tag{30}$$

In Eq. (29), the integration with respect to ϕ is carried out. The variable of integration is changed in accordance with

$$p^2/k^2 = 1 + \tau^2 \tag{31}$$

Then, P_{im} simplifies as

$$P_{im} = -k^2 w_0^2 \exp\left[-\frac{k^2 w_0^2}{2} \left(1 + \frac{b_t}{b} \right) \right]$$
$$\times \int_0^\infty d\tau (1 - \tau^2) \exp\left[-\frac{k^2 w_0^2}{2} \left(1 - \frac{b_t}{b} \right) \tau^2 \right] \tag{32}$$

The integral is evaluated, with the result that

$$P_{im} = -\left(\frac{\pi}{2}\right)^{1/2} kw_0 \left(1 - \frac{b_t}{b}\right)^{-1/2} \exp\left[-\frac{k^2 w_0^2}{2}\left(1 + \frac{b_t}{b}\right)\right]$$
$$\times \left[1 - \frac{1}{k^2 w_0^2 (1 - b_t/b)}\right] \tag{33}$$

An explicit expression has been obtained for the reactive power in terms of the parameters kw_0 and b_t/b. For $b_t/b = 0$, Eq. (33) becomes the same as Eq. (2.36); that is, Eq. (33) reproduces correctly the reactive power of the fundamental Gaussian wave for the limiting case of $b_t/b = 0$. For $b_t/b = 1$, Eq. (33) shows that the reactive power is infinite [see Eq. (4.25)]. Thus, Eq. (33) also reproduces correctly the reactive power of the basic full Gaussian wave for the other limiting case of $b_t/b = 1$. For a certain combination of the parameters kw_0 and b_t/b, there is resonance in that the reactive power is zero. For suitable changes in kw_0 and b_t/b, the reactive power changes from inductive to capacitive or vice versa.

The reactive power P_{im} is shown in Fig. 5 as a function of b_t/b for $0 \le b_t/b \le 1$ and $kw_0 = 1.563$. The real power decreases as b_t/b is increased from 0 to 1 (see Fig. 1). But the reactive power increases as b_t/b is increased from 0 to 1. For $kw_0 = 1.563$, the extended full Gaussian wave has a resonance in that $P_{im} = 0$ for $b_t/b = 0.5905$. As b_t/b comes closer to 1, P_{im} increases rapidly as b_t/b approaches 1 and becomes infinite for $b_t/b = 1$. In Fig. 6, the reactive power is shown as a function of kw_0 for $0.2 < kw_0 < 1.2$ and $b_t/b = 0.5$. For $b_t/b < 1$, the reactive power decreases and approaches zero as kw_0 is increased.

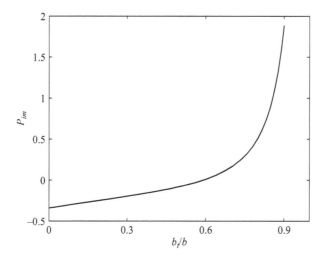

Fig. 5. Reactive power P_{im} of the extended full Gaussian wave as a function of b_t/b for $0 < b_t/b < 1$ and $kw_0 = 1.563$. $P_{im} = 0$ occurs at $b_t/b = 0.5905$. Time-averaged power in the corresponding paraxial beam is given by $P_0^\pm = 1$ W.

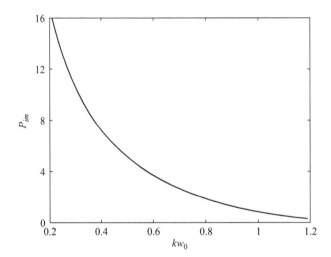

Fig. 6.   Reactive power P_{im} of the extended full Gaussian wave as a function of kw_0 for $0.2 < kw_0 < 1.2$ and $b_t/b = 0.5$. Time-averaged power in the corresponding paraxial beam is given by $P_0^{\pm} = 1$ W.

The paraxial approximation corresponds to $kw_0 \to \infty$. In the paraxial approximation, the reactive power $P_{im} = 0$. Figure 6 reveals that the paraxial beam result is reached for kw_0 as small as 1.2.

References

1. S. R. Seshadri, "Reactive power in the full Gaussian light wave," *J. Opt. Soc. Am. A* **26**, 2427–2433 (2009).

2. S. R. Seshadri, "Full-wave generalizations of the fundamental Gaussian beam," *J. Opt. Soc. Am. A* **26**, 2515–2520 (2009).

3. S. R. Seshadri, "Dynamics of the linearly polarized fundamental Gaussian light wave," *J. Opt. Soc. Am. A* **24**, 482–492 (2007).

4. S. R. Seshadri, "Quality of paraxial electromagnetic beams," *Appl. Opt.* **45**, 5335–5345 (2006).

Cylindrically symmetric transverse magnetic full Gaussian wave

The fundamental Gaussian beam and wave are generated by the magnetic/electric vector potential, oriented perpendicular to the direction of propagation. If the vector potential is oriented parallel to the propagation direction and is cylindrically symmetrical, transverse magnetic (TM) and transverse electric (TE) beams and waves are excited [1–5]. The lowest-order solution is used for the vector potential and is the same as for the fundamental Gaussian beam and wave. The resulting electromagnetic fields are one order higher, which is one order of magnitude in kw_0 smaller, than those for the fundamental Gaussian beam. For the TM paraxial beam, the electric field component $E_\rho(\rho, z)$ is discontinuous across the secondary source plane $z = 0$, resulting in an azimuthally directed magnetic current sheet on the secondary source plane. Therefore, the problem of the cylindrically symmetric TM Gaussian paraxial beam and full wave is formulated directly in terms of the electromagnetic fields and the source current density. The solution is first obtained in the paraxial approximation. The electromagnetic fields are determined and the characteristics of the real power, the radiation intensity distribution, and the reactive power are obtained. The reactive power of the TM Gaussian beam vanishes.

The full-wave generalization of the cylindrically symmetric TM paraxial beam is carried out. The procedure involves the analytic continuation of the asymptotic $(|z| \to \infty)$ field from $|z|$ to $|z| - ib_t$, where the length parameter b_t is in the range $0 \le b_t \le b$ and b is the Rayleigh distance. The limiting case of $b_t/b = 1$ is identical to the full-wave treatment introduced by Deschamps [6] and Felsen [7] for the fundamental Gaussian beam [4]. For the cylindrically symmetric TM full Gaussian wave, the required magnetic current density is deduced and the relevant components of the generated electromagnetic fields are determined. The characteristics of the real power, the reactive power, and the radiation intensity distribution are analyzed. The reactive power of the cylindrically symmetric TM full Gaussian wave does not vanish. The dependence of the characteristics of the various physical quantities on b_t/b and kw_0 is examined.

1 Cylindrically symmetric transverse magnetic beam

The outward propagations in the $+z$ direction in $z > 0$ and in the $-z$ direction in $z < 0$ are considered. The secondary source is an infinitesimally thin sheet of current located on the

$z = 0$ plane. The current does not vary in the ϕ (azimuthal) direction; therefore, all the excited fields are independent of ϕ. The currents are assumed to be in the ϕ direction. The magnetic and electric current densities are denoted by $\hat{\phi} J_m(\rho, z)$ and $\hat{\phi} J_e(\rho, z)$, respectively. The components of Maxwell's equations [Eqs. (D8) and (D9)] are grouped into two independent sets as follows:

$$ikH_\phi - J_m = \frac{\partial E_\rho}{\partial z} - \frac{\partial E_z}{\partial \rho} \tag{1}$$

$$-ikE_\rho = -\frac{\partial H_\phi}{\partial z} \tag{2}$$

$$-ikE_z = \frac{1}{\rho}\frac{\partial(\rho H_\phi)}{\partial \rho} = \left(\frac{\partial}{\partial \rho} + \frac{1}{\rho}\right) H_\phi \tag{3}$$

$$-ikE_\phi + J_e = \frac{\partial H_\rho}{\partial z} - \frac{\partial H_z}{\partial \rho} \tag{4}$$

$$ikH_\rho = -\frac{\partial E_\phi}{\partial z} \tag{5}$$

$$ikH_z = \frac{1}{\rho}\frac{\partial(\rho E_\phi)}{\partial \rho} = \left(\frac{\partial}{\partial \rho} + \frac{1}{\rho}\right) E_\phi \tag{6}$$

The first set of field components, (H_ϕ, E_ρ, E_z), given by Eqs. (1)–(3), is excited by the magnetic current of density J_m. This set forms the TM mode since the magnetic field H_ϕ is perpendicular to the direction $(\pm z)$ of propagation. The polarization is stated with reference to the transverse electric field. Therefore, the TM mode is radially polarized. The second set of field components, (E_ϕ, H_ρ, H_z), given by Eqs. (4)–(6), is generated by the electric current of density J_e. This set forms the TE mode since the electric field E_ϕ is perpendicular to the direction of propagation. The TE mode is azimuthally polarized. The second set is obtained from the first by the use of the following duality relations:

$$E_\rho \rightarrow H_\rho, \quad E_z \rightarrow H_z, \quad H_\phi \rightarrow -E_\phi, \quad J_m \rightarrow -J_e \tag{7}$$

Since the second set is the dual to the first, only the TM mode with radial polarization is treated.

When E_ρ and E_z from Eqs. (2) and (3), respectively, are substituted into Eq. (1), the differential equation satisfied by H_ϕ is found as

$$\left(\frac{\partial^2}{\partial \rho^2} + \frac{1}{\rho}\frac{\partial}{\partial \rho} - \frac{1}{\rho^2} + \frac{\partial^2}{\partial z^2} + k^2\right) H_\phi(\rho, z) = -ikJ_m(\rho, z) \tag{8}$$

The current density is specified as

$$J_m(\rho, z) = J_m(\rho)\delta(z) \tag{9}$$

It follows from Eqs. (8) and (9) that $H_\phi(\rho, z)$ is continuous across the $z = 0$ plane. The azimuthally directed current cannot exist at $\rho = 0$; that is, $J_m(\rho)$ is required to vanish at

$\rho = 0$. As a consequence, $H_\phi(\rho, 0)$ should vanish for $\rho = 0$. Therefore, a simple form is assumed for the input $(z = 0)$ value of $H_\phi(\rho, z)$ as

$$H_{\phi 0}(\rho, 0) = N\rho \exp\left(-\rho^2/w_0^2\right) \tag{10}$$

where w_0 is the scale length of variation of the fields in the radial direction and N is a normalization constant to be determined at a later stage in the analysis. The additional subscript 0 is used to denote paraxial. In view of Eq. (9), it follows that the right side of Eq. (8) vanishes for $|z| > 0$. We shall obtain the propagation characteristics of the TM paraxial beam for $|z| > 0$ starting from Eqs. (8) and (10).

The plane-wave phase factor is separated out to obtain

$$H_{\phi 0}^{\pm}(\rho, z) = \exp(\pm ikz)h^{\pm}(\rho, z) \tag{11}$$

where the sign $\pm$ and the superscript $\pm$ denote the propagation in the $\pm z$ direction and $h^{\pm}(\rho, z)$ is the slowly varying amplitude. Equation (11) is substituted into Eq. (8) and the paraxial approximation (see Appendix C.1) is made to obtain the paraxial wave equation satisfied by $h^{\pm}(\rho, z)$ as

$$\left(\frac{\partial^2}{\partial\rho^2} + \frac{1}{\rho}\frac{\partial}{\partial\rho} - \frac{1}{\rho^2} \pm 2ik\frac{\partial}{\partial z}\right)h^{\pm}(\rho, z) = 0 \tag{12}$$

Since

$$\left(\frac{\partial^2}{\partial\rho^2} + \frac{1}{\rho}\frac{\partial}{\partial\rho} + \eta^2 - \frac{1}{\rho^2}\right)J_1(\eta\rho) = 0 \tag{13}$$

where $J_1(\ )$ is the Bessel function of order 1, the following Bessel transform pair is used for the solution of Eq. (12):

$$h^{\pm}(\rho, z) = \int_0^\infty d\eta\eta J_1(\eta\rho)\bar{h}^{\pm}(\eta, z) \tag{14}$$

$$\bar{h}^{\pm}(\eta, z) = \int_0^\infty d\rho\rho J_1(\eta\rho)h^{\pm}(\rho, z) \tag{15}$$

When Eq. (14) is substituted into Eq. (12), it is found that

$$\left(\frac{\partial}{\partial z} \pm \frac{i\eta^2 w_0^2}{4b}\right)\bar{h}^{\pm}(\eta, z) = 0 \tag{16}$$

where $b = \frac{1}{2}kw_0^2$ is the Rayleigh distance. The solution of Eq. (16) is given by

$$\bar{h}^{\pm}(\eta, z) = \bar{h}^{\pm}(\eta, 0)\exp\left(-\frac{\eta^2 w_0^2}{4}\frac{i|z|}{b}\right) \tag{17}$$

From Eqs. (10) and (15), the Bessel transform of $h^\pm(\rho, 0)$ $[= H_{\phi 0}(\rho, 0)]$ is found as

$$\bar{h}^\pm(\eta, 0) = N \int_0^\infty d\rho \rho J_1(\eta \rho) \rho \exp(-\rho^2/w_0^2) \tag{18}$$

The use of the integral relation [8]

$$\int_0^\infty dt t^{n+1} J_n(at) \exp(-p^2 t^2) = \frac{a^n}{(2p^2)^{n+1}} \exp\left(-\frac{a^2}{4p^2}\right) \tag{19}$$

enables us to find $\bar{h}^\pm(\eta, 0)$ as

$$\bar{h}^\pm(\eta, 0) = N \frac{w_0^4}{4} \eta \exp\left(-\frac{\eta^2 w_0^2}{4}\right) \tag{20}$$

Substituting Eq. (20) into Eq. (17) and carrying out the inverse Bessel transform by the use of Eq. (14) yields

$$h^\pm(\rho, z) = N \frac{w_0^4}{4} \int_0^\infty d\eta \eta^2 J_1(\eta \rho) \exp\left(-\frac{\eta^2 w_0^2}{4 q_\pm^2}\right) \tag{21}$$

where

$$q_\pm = (1 + i|z|/b)^{-1/2} \tag{22}$$

For the position coordinates in the physical space, $1/q_\pm^2 \neq 0$, and Eq. (19) is used to find $h^\pm(\rho, z)$ from Eq. (21) as

$$h^\pm(\rho, z) = N q_\pm^4 \rho \exp(-q_\pm^2 \rho^2/w_0^2) \tag{23}$$

From Eqs. (11) and (23), it is found that

$$H_{\phi 0}^\pm(\rho, z) = \exp(\pm ikz) N q_\pm^4 \rho \exp(-q_\pm^2 \rho^2/w_0^2) \tag{24}$$

which is the field of the cylindrically symmetric TM paraxial beam. From Eqs. (2) and (24), in the paraxial approximation, $E_\rho^\pm(\rho, z)$ is determined as

$$E_{\rho 0}^\pm(\rho, z) = \pm \exp(\pm ikz) N q_\pm^4 \rho \exp(-q_\pm^2 \rho^2/w_0^2) \tag{25}$$

To find the time-averaged Poynting vector and the complex power, only the transverse field components, $H_{\phi 0}^\pm(\rho, z)$ and $E_{\rho 0}^\pm(\rho, z)$, are required; the only other field component, namely the longitudinal field component, $E_{z0}^\pm(\rho, z)$, is not needed.

The z component of the time-averaged Poynting vector $S_{z0}^\pm(\rho, z)$ is derived from Eqs. (24) and (25). By integrating $[\pm S_{z0}^\pm(\rho, z)]$ throughout the entire transverse plane, the time-averaged power $P_0^\pm$ transported by the TM paraxial beam in the $+z$ and $-z$ directions is determined as

$$P_0^{\pm} = 2\pi \int_0^{\infty} d\rho \rho \left(\pm \frac{c}{2}\right) \text{Re}\left[E_{\rho 0}^{\pm}(\rho, z) H_{\phi 0}^{\pm *}(\rho, z)\right]$$

$$= \frac{cN^2\pi}{(1 + z^2/b^2)^2} \int_0^{\infty} d\rho \rho^3 \exp\left[-\frac{2\rho^2}{w_0^2(1 + z^2/b^2)}\right]$$

$$= \frac{cN^2\pi w_0^4}{8} \tag{26}$$

The normalization constant is chosen as

$$N = \left(8/c\pi w_0^4\right)^{1/2} \tag{27}$$

Then, $P_0^{\pm} = 1$ W. The normalization is such that the time-averaged power transported by the TM paraxial beam in the $\pm z$ direction is given by $P_0^{\pm} = 1$ W.

For $z > 0$, the z component of the time-averaged Poynting vector $S_{z0}^+(\rho, z)$ is found and the expression for N^2 is substituted from Eq. (27), with the result that

$$S_{z0}^+(\rho, z) = \frac{4}{\pi w_0^4} \frac{\rho^2}{\left(1 + z^2/b^2\right)^2} \exp\left[-\frac{2\rho^2}{w_0^2\left(1 + z^2/b^2\right)}\right] \tag{28}$$

The spherical coordinates (r, θ^+, ϕ) are used. Then, $\rho = r \sin \theta^+$ and $z = r \cos \theta^+$, where θ^+ is defined with respect to the $+z$ axis. Since $\hat{z} = \hat{r} \cos \theta^+ - \hat{\theta} \sin \theta^+$, the radial component of the time-averaged Poynting vector is given by $S_{r0}^+(r, \theta^+, \phi) = \cos \theta^+ S_{z0}^+(\rho, z)$. The radiation intensity for $z > 0$ is then determined as

$$\Phi_0(\theta^+, \phi) = \lim_{kr \to \infty} r^2 S_{r0}^+(r, \theta^+, \phi)$$

$$= \frac{k^4 w_0^4}{4\pi} \frac{\tan^2 \theta^+}{\cos \theta^+} \exp\left(-\frac{1}{2} k^2 w_0^2 \tan^2 \theta^+\right) \tag{29}$$

For $z < 0$, $S_{z0}^-(\rho, z)$ corresponding to Eq. (28) is found as

$$S_{z0}^-(\rho, z) = -S_{z0}^+(\rho, z) \tag{30}$$

The spherical coordinates (r, θ^-, ϕ) are used. Here, θ^- is defined with respect to the $-z$ axis. Therefore, $\rho = r \sin \theta^-$ and $-z = r \cos \theta^-$. Also, since $-\hat{z} = \hat{r} \cos \theta^- - \hat{\theta} \sin \theta^-$, the radial component of the time-averaged Poynting vector is found as $S_{r0}^-(r, \theta^-, \phi) = -\cos \theta^- S_{z0}^-(\rho, z)$ The radiation intensity for $z < 0$ is then evaluated as

$$\Phi_0(\theta^-, \phi) = \lim_{kr \to \infty} r^2 S_{r0}^-(r, \theta^-, \phi)$$

$$= \frac{k^4 w_0^4}{4\pi} \frac{\tan^2 \theta^-}{\cos \theta^-} \exp\left(-\frac{1}{2} k^2 w_0^2 \tan^2 \theta^-\right) \tag{31}$$

The radiation intensity distribution $\Phi_0(\theta^-, \phi)$ for $z < 0$ is found from the corresponding distribution $\Phi_0(\theta^+, \phi)$ for $z > 0$ by reflection about the $z = 0$ plane.

From Eqs. (24) and (25), it is seen that $H_{\phi 0}^{\pm}(\rho, z)$ is continuous and $E_{\rho 0}^{\pm}(\rho, z)$ is discontinuous across the plane $z = 0$. The discontinuity of the tangential electric field is equivalent to a magnetic current density induced on the secondary source plane. The induced magnetic current density is deduced from Eq. (25) as

$$\mathbf{J}_{m0}(\rho, z) = -\hat{z} \times \hat{\rho}[E_\rho^+(\rho, 0) - E_\rho^-(\rho, 0)]\delta(z)$$
$$= -\hat{\phi}2N\rho \exp(-\rho^2/w_0^2)\delta(z) \tag{32}$$

The complex power is determined from Eqs. (24), (32), and (D22) as

$$P_C = -\frac{c}{2}2\pi \int_0^\infty d\rho\rho \int_{-\infty}^\infty dz \mathbf{H}_0^\pm(\rho, z) \cdot \mathbf{J}_{m0}^*(\rho, z)$$
$$= c\pi 2 \int_0^\infty d\rho\rho N^2 \rho^2 \exp\left(-\frac{2\rho^2}{w_0^2}\right) = 2 \text{ W} \tag{33}$$

The value of N^2 is substituted from Eq. (27) in obtaining Eq. (33). The circularly symmetric TM paraxial beam has no reactive power; that is, $P_{im} = 0$. The real power $P_{re} = P_0^+ + P_0^- = 2$ W, of which 1 W flows in the $+z$ direction and the other 1 W flows in the $-z$ direction.

The reactive power of the fundamental Gaussian beam is found to be zero. Here, we find that the reactive power of the cylindrically symmetric TM paraxial beam is also zero. As for the fundamental Gaussian beam, and for the cylindrically symmetric TM paraxial beam as well, there is a need for the full-wave generalization of the paraxial beam.

2 Current source in complex space

For the position coordinates in the physical space, Eq. (21) yielded the conventional expression [Eq. (24)] for the cylindrically symmetric TM beam. The domain of z is extended to include complex values. The value of $h^\pm(\rho, z)$ is sought for $|z| - ib = 0$. Then, $1/q_\pm^2 = 0$, and Eq. (21) leads to the source in the complex space as follows:

$$C_{s0}(\rho, z) = N\frac{w_0^4}{4}\int_0^\infty d\eta \eta^2 J_1(\eta\rho) \quad \text{for } |z| - ib = 0 \tag{34}$$

Since $J_1(\eta\rho) = -J_0'(\eta\rho)$, where prime denotes differentiation with respect to the argument and $(\partial/\partial\rho)J_0(\eta\rho) = \eta J_0'(\eta\rho)$, Eq. (34) is recast as

$$C_{s0}(\rho, z) = -N\frac{w_0^4}{4}\frac{\partial}{\partial\rho}\int_0^\infty d\eta \eta J_0(\eta\rho) \quad \text{for } |z| - ib = 0 \tag{35}$$

Consider the following Bessel transform pair similar to Eqs. (14) and (15) but with $J_1(\eta\rho)$ replaced by $J_0(\eta\rho)$:

$$f(\rho, z) = \int_0^\infty d\eta \eta J_0(\eta\rho)\overline{f}(\eta, z) \tag{36}$$

$$\overline{f}(\eta, z) = \int_0^\infty d\rho\rho J_0(\eta\rho)f(\rho, z) \tag{37}$$

Let $f(\rho, z) = \delta(\rho)/\rho$. Since $J_0(0) = 1$, it follows from Eq. (37) that $\bar{f}(\eta, z) = 1$. Then, from Eq. (36), it is found that

$$\int_0^\infty d\eta\,\eta J_0(\eta\rho) = \frac{\delta(\rho)}{\rho} \tag{38}$$

Substituting Eq. (38) into Eq. (35) yields

$$C_{s0}(\rho, z) = -N \frac{w_0^4}{4} \frac{\partial}{\partial \rho} \frac{\delta(\rho)}{\rho} \quad \text{for } |z| - ib = 0 \tag{39}$$

Thus, the source in the complex space corresponding to $|z| - ib = 0$ is a higher-order point source. For the fundamental Gaussian beam, the corresponding source [see Eq. (5.5)] in the complex space is a point source, but without the derivative. Since the source in the complex space is a higher-order point source, as for the basic full Gaussian wave, and for the full-wave generalization of the cylindrically symmetric TM beam as well, the reactive power is expected to be infinite. In order to obtain finite reactive power, it is necessary to look for a current distribution rather than a higher-order point source in the complex space.

The search for the desired complex space source is carried out at $|z| - ib_t = 0$, where b_t is real. Then, $1/q_\pm^2 = 1 - b_t/b$. It is assumed that $0 \leq b_t/b \leq 1$. Equations (19) and (21) yield the source in the complex space as

$$C_{s0}(\rho, z) = N \frac{w_0^4}{4} \int_0^\infty d\eta\,\eta^2 J_1(\eta\rho) \exp\left[-\frac{\eta^2 w_0^2}{4}\left(1 - \frac{b_t}{b}\right)\right] \tag{40}$$

$$= N \frac{\rho}{(1 - b_t/b)^2} \exp\left[-\frac{\rho^2}{w_0^2(1 - b_t/b)}\right] \tag{41}$$

The source given by Eq. (40) or (41), located at $|z| - ib_t = 0$, generates the paraxial beam for all $|z| > 0$. For $|z| \to \infty$, the source moves to $|z| = 0$. Therefore, the source is moved to $|z| = 0$ for the determination of the asymptotic value of the paraxial beam. The effect of moving the source back to $|z| - ib_t = 0$ is realized by the analytic continuation of the asymptotic value of the paraxial beam from $|z|$ to $|z| - ib_t$. For the paraxial beam the asymptotic value is found only for the slowly varying amplitude. But the analytic continuation has to be carried out on the total field, which also includes the rapidly varying plane-wave phase: $\exp(\pm ikz)$. The paraxial wave equation is only an approximation for the full Helmholtz wave equation. Therefore, the complex space source deduced from the paraxial beam, except for an excitation coefficient S_{ex}, is used for the Helmholtz equation to obtain the asymptotic value of the full Gaussian wave. Hence, the complex space source for obtaining the full-wave generalization of the TM beam is obtained from Eq. (40) as

$$C_s(\rho, z) = S_{ex} N \frac{w_0^4}{4} \int_0^\infty d\eta\,\eta^2 J_1(\eta\rho) \exp\left[-\frac{\eta^2 w_0^2}{4}\left(1 - \frac{b_t}{b}\right)\right] \tag{42}$$

The excitation coefficient S_{ex} is found from the requirement in the paraxial approximation that the full wave should correctly reproduce the initially chosen paraxial beam.

The source that yields the asymptotic value of the full-wave solution of the Helmholtz equation is found from Eq. (42) as

$$C_{s\infty}(\rho,z) = S_{ex}N\frac{w_0^4}{4}\delta(z)\int_0^\infty d\eta\eta^2 J_1(\eta\rho)\exp\left[-\frac{\eta^2 w_0^2}{4}\left(1 - \frac{b_t}{b}\right)\right] \tag{43}$$

3 Cylindrically symmetric TM full wave

Let $G(\rho,z)$ be the solution of the Helmholtz equation for the source given by Eq. (43). Then, $G(\rho,z)$ satisfies the following inhomogeneous Helmholtz equation:

$$\left(\frac{\partial^2}{\partial\rho^2} + \frac{1}{\rho}\frac{\partial}{\partial\rho} - \frac{1}{\rho^2} + \frac{\partial^2}{\partial z^2} + k^2\right)G(\rho,z) = -S_{ex}N\frac{w_0^4}{4}\delta(z)\int_0^\infty d\eta\eta^2 J_1(\eta\rho)$$

$$\times \exp\left[-\frac{\eta^2 w_0^2}{4}\left(1 - \frac{b_t}{b}\right)\right] \tag{44}$$

$G(\rho,z)$ is replaced by its Bessel transform representation, similar to that given by Eq. (14). Then, it is found that $\overline{G}(\eta,z)$, the Bessel transform of $G(\rho,z)$, satisfies the following differential equation:

$$\left(\frac{\partial^2}{\partial z^2} + \zeta^2\right)\overline{G}(\eta,z) = -S_{ex}N\frac{w_0^4}{4}\eta\exp\left[-\frac{\eta^2 w_0^2}{4}\left(1 - \frac{b_t}{b}\right)\right]\delta(z) \tag{45}$$

where

$$\zeta = \left(k^2 - \eta^2\right)^{1/2} \quad \text{for } k^2 > \eta^2$$
$$= i\left(\eta^2 - k^2\right)^{1/2} \quad \text{for } k^2 < \eta^2 \tag{46}$$

The solution of Eq. (45) is determined as

$$\overline{G}(\eta,z) = iS_{ex}N\frac{w_0^4}{8}\eta\exp\left[-\frac{\eta^2 w_0^2}{4}\left(1 - \frac{b_t}{b}\right)\right]$$

$$\times \zeta^{-1}\exp(i\zeta|z|) \tag{47}$$

$\overline{G}(\eta,z)$ is analytically continued from $|z|$ to $|z| - ib_t$ to find the Bessel transform of the magnetic field as

$$\overline{H}_\phi^\pm(\eta,z) = iS_{ex}N\frac{w_0^4}{8}\eta\exp\left[-\frac{\eta^2 w_0^2}{4}\left(1 - \frac{b_t}{b}\right)\right]$$

$$\times \zeta^{-1}\exp[i\zeta(|z| - ib_t)] \tag{48}$$

The paraxial approximation corresponds to $\eta^2 \ll k^2$. Then, as given by Eq. (C29), ζ is expanded as

$$\zeta = k - \eta^2 w_0^2/4b \tag{49}$$

In Eq. (48), if ζ in the amplitude is replaced by the first term in Eq. (49) and ζ in the phase is replaced by the first two terms in Eq. (49), the paraxial approximation of Eq. (48) is obtained as

$$\overline{H}_\phi^\pm(\eta, z) = \exp(\pm ikz)iS_{ex}N\frac{w_0^4}{8k}\exp(kb_t)\eta \exp\left[-\frac{\eta^2 w_0^2}{4q_\pm^2}\right] \tag{50}$$

The inverse Bessel transform of Eq. (50) is

$$H_\phi^\pm(\rho, z) = \exp(\pm ikz)\frac{iS_{ex}}{2k}\exp(kb_t)Nq_\pm^4\rho \exp\left(-\frac{q_\pm^2\rho^2}{w_0^2}\right) \tag{51}$$

The excitation coefficient is chosen as

$$S_{ex} = -i2k\exp(-kb_t) \tag{52}$$

Then, a comparison of Eq. (51) with Eq. (24) shows that the paraxial approximation of Eq. (48) correctly reproduces the initially chosen paraxial beam. Substituting S_{ex} from Eq. (52) into Eq. (48) and taking the inverse Bessel transform yields

$$H_\phi^\pm(\rho, z) = k\exp(-kb_t)N\frac{w_0^4}{4}\int_0^\infty d\eta\eta^2 J_1(\eta\rho)$$

$$\times \exp\left[-\frac{\eta^2 w_0^2}{4}\left(1 - \frac{b_t}{b}\right)\right]\zeta^{-1}\exp[i\zeta(|z| - ib_t)] \tag{53}$$

The full-wave generalization of the cylindrically symmetric TM paraxial beam is given by Eq. (53) and is designated the cylindrically symmetric TM full Gaussian wave.

4 Real power

From Eqs. (2) and (53), for the cylindrically symmetric TM full wave, the transverse electric field is obtained as

$$E_\rho^\pm(\rho, z) = \frac{1}{ik}\frac{\partial H_\phi^\pm(\rho, z)}{\partial z} = \pm\exp(-kb_t)N\frac{w_0^4}{4}\int_0^\infty d\overline{\eta}\overline{\eta}^2 J_1(\overline{\eta}\rho)$$

$$\times \exp\left[-\frac{\overline{\eta}^2 w_0^2}{4}\left(1 - \frac{b_t}{b}\right)\right]\exp\left[i\overline{\zeta}(|z| - ib_t)\right] \tag{54}$$

where $\bar{\zeta}$ is the same as ζ, with η replaced by $\bar{\eta}$. The discontinuity of $E_\rho^\pm(\rho, z)$ across the plane $z = 0$ is equivalent to a magnetic current density induced on the secondary source plane. The induced magnetic current density is obtained from Eq. (54) as

$$\mathbf{J}_m(\rho, z) = -\hat{z} \times \hat{\rho} [E_\rho^+(\rho, 0) - E_\rho^-(\rho, 0)]\delta(z)$$

$$= -\hat{\phi} 2 \exp(-kb_t) N \frac{w_0^4}{4} \delta(z) \int_0^\infty d\bar{\eta}\bar{\eta}^2 J_1(\bar{\eta}\rho)$$

$$\times \exp\left[-\frac{\bar{\eta}^2 w_0^2}{4}\left(1 - \frac{b_t}{b}\right)\right] \exp\left(\bar{\zeta} b_t\right) \tag{55}$$

The complex power is determined from Eqs. (53), (55), and (D22) as

$$P_C = -\frac{c}{2} \int_0^\infty d\rho\rho \int_0^{2\pi} d\phi \int_{-\infty}^\infty dz \mathbf{H}^\pm(\rho, z) \cdot \mathbf{J}_m^{\pm *}(\rho, z)$$

$$= -\frac{c}{2} \int_0^\infty d\rho\rho \int_0^{2\pi} d\phi k \exp(-kb_t) N \frac{w_0^4}{4} \int_0^\infty d\eta\eta^2 J_1(\eta\rho)$$

$$\times \exp\left[-\frac{\eta^2 w_0^2}{4}\left(1 - \frac{b_t}{b}\right)\right] \zeta^{-1} \exp(\zeta b_t)$$

$$\times (-2)\exp(-kb_t) N \frac{w_0^4}{4} \int_0^\infty d\bar{\eta}\bar{\eta}^2 J_1(\bar{\eta}\rho)$$

$$\times \exp\left[-\frac{\bar{\eta}^2 w_0^2}{4}\left(1 - \frac{b_t}{b}\right)\right] \exp(\bar{\zeta}^* b_t) \tag{56}$$

In Eq. (14), let $\bar{h}^\pm(\eta, z) = \delta(\eta - \bar{\eta})/\eta$; then, $h^\pm(\rho, z) = J_1(\bar{\eta}\rho)$. Substituting $\bar{h}^\pm(\eta, z)$ and $h^\pm(\rho, z)$ into Eq. (15) yields

$$\frac{\delta(\eta - \bar{\eta})}{\eta} = \int_0^\infty d\rho\rho J_1(\eta\rho)J_1(\bar{\eta}\rho) \tag{57}$$

The ρ integration in Eq. (56) is carried out by the use of Eq. (57). Then, the $\bar{\eta}$ integration is performed with the result that

$$P_C = \frac{c}{2} k \exp(-2kb_t) \frac{N^2 w_0^8}{8} \int_0^{2\pi} d\phi \int_0^\infty d\eta\eta^3$$

$$\times \exp\left[-\frac{\eta^2 w_0^2}{2}\left(1 - \frac{b_t}{b}\right)\right] \zeta^{-1} \exp[b_t(\zeta + \zeta^*)] \tag{58}$$

For $\eta < k$, $\zeta = (k^2 - \eta^2)^{1/2}$ is real and the integrand is real. For $\eta > k$, $\zeta = i(\eta^2 - k^2)^{1/2}$ is imaginary and the integrand is imaginary. Therefore, P_{re}, the real part of P_C, is found from Eq. (58) as

$$P_{re} = k \exp(-2kb_t) \frac{w_0^4}{2\pi} \int_0^{2\pi} d\phi \int_0^k d\eta \eta^3 \exp\left[-\frac{\eta^2 w_0^2}{2}\left(1 - \frac{b_t}{b}\right)\right] \zeta^{-1} \exp(2b_t\zeta) \tag{59}$$

The value of N^2 from Eq. (27) is substituted in obtaining Eq. (59). The variable is changed to yield

$$\eta = k \sin \theta^+, \qquad \zeta = k|\cos \theta^+| \tag{60}$$

For $\eta = 0$, $\theta^+ = 0$ or π, and for $\eta = k$, $\theta^+ = \pi/2$. Therefore, P_{re} can be transformed as

$$P_{re} = \frac{k^4 w_0^4}{4\pi} \int_0^{2\pi} d\phi \left(\int_0^{\pi/2} + \int_\pi^{\pi/2}\right) d\theta^+ \cos \theta^+ \sin^3 \theta^+$$

$$\times \exp\left[-\frac{k^2 w_0^2}{2}\left(1 - \frac{b_t}{b}\right) \sin^2 \theta^+\right] \frac{\exp[-2kb_t(1 - |\cos \theta^+|)]}{|\cos \theta^+|} \tag{61}$$

For $0 < \theta^+ < \pi/2$, $|\cos \theta^+| = \cos \theta^+$. For $\pi > \theta^+ > \pi/2$, $|\cos \theta^+| = -\cos \theta^+$. Let

$$\theta^- = \pi - \theta^+ \qquad \text{for } \pi > \theta^+ > \pi/2 \tag{62}$$

Then, $\pi > \theta^+ > \pi/2$ changes to $0 < \theta^- < \pi/2$ and $\cos \theta^+ = -\cos \theta^-$. Consequently, P_{re} in Eq. (61) is expressed as

$$P_{re} = P_{TM}^+ + P_{TM}^- \tag{63}$$

where

$$P_{TM}^\pm = \int_0^{\pi/2} d\theta^\pm \sin \theta^\pm \int_0^{2\pi} d\phi \Phi_{TM}(\theta^\pm, \phi) \tag{64}$$

and

$$\Phi_{TM}(\theta^\pm, \phi) = \frac{k^4 w_0^4}{4\pi} \sin^2 \theta^\pm \exp\left[-\frac{k^2 w_0^2}{2}\left(1 - \frac{b_t}{b}\right) \sin^2 \theta^\pm\right]$$

$$\times \exp\left[-k^2 w_0^2 \frac{b_t}{b}(1 - \cos \theta^\pm)\right] \tag{65}$$

The real power P_{re} is the time-averaged power created by the magnetic current source. The time-averaged powers transported by the cylindrically symmetric TM full wave in the $+z$ and $-z$ directions are given by P_{TM}^+ and P_{TM}^-, respectively. From Eqs. (64) and (65), it follows that $P_{TM}^+ = P_{TM}^-$. The normalization is such that the time-averaged power carried by the corresponding paraxial beam in the $\pm z$ direction is given by $P_0^\pm = 1$ W.

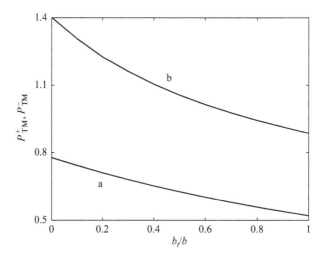

Fig. 1. Time-averaged power $P_{TM}^{\pm}$ transported by the cylindrically symmetric TM full Gaussian wave in the $\pm z$ direction as a function of b_t/b for $0 < b_t/b < 1$, and (a) $kw_0 = 1.563$ and (b) $kw_0 = 2.980$. Time-averaged power $P_0^{\pm}$ carried by the corresponding paraxial beam in the $\pm z$ direction is $P_0^{\pm} = 1$ W.

In Eq. (64), the ϕ integration is carried out analytically and the $\theta^{\pm}$ integration is performed numerically to obtain $P_{TM}^{\pm}$ as functions of b_t/b and kw_0. In Fig. 1, $P_{TM}^{\pm}$ is shown as a function of b_t/b for $0 \le b_t/b \le 1$, and (a) $kw_0 = 1.563$ and (b) $kw_0 = 2.980$. $P_{TM}^{\pm}$ is observed to decrease monotonically as b_t/b is increased from 0 to 1. For a fixed b_t/b, $P_{TM}^{\pm}$ is generally larger for the larger value of kw_0. In Fig. 2, $P_{TM}^{\pm}$ is depicted as a function of kw_0 for $1 < kw_0 < 10$ and for (a) $b_t/b = 0$, (b) $b_t/b = 0.5$, and (c) $b_t/b = 1$. The curves for all b_t/b have a low-frequency region where $P_{TM}^{\pm}$ increases as kw_0 is increased and a high-frequency or paraxial region where $P_{TM}^{\pm}$ approaches $P_0^{\pm} = 1$ W as kw_0 is increased. For $b_t/b = 0$, the curve of $P_{TM}^{\pm}$ has a resonant peak, with the peak value greater than $P_0^{\pm} = 1$ W and located between the low- and the high-frequency regions. As b_t/b is increased, the resonant peak decreases and disappears, and $P_{TM}^{\pm}$ increases monotonically, approaching $P_0^{\pm} = 1$ W as kw_0 is increased. For a fixed b_t/b, except for a small range of kw_0, $P_{TM}^{\pm}/P_0^{\pm}$ approaches $P_0^{\pm} = 1$ W as kw_0 is increased. Consequently, in general, the quality of the paraxial approximation to the full wave for a fixed b_t/b improves as kw_0 is increased [4].

5 Reactive power

For $k < \eta < \infty$, ζ is imaginary, and in Eq. (58), the integrand is imaginary. Therefore, the reactive power P_{im}, the imaginary part of P_C, is found from Eq. (58) as

$$P_{im} = -k \exp(-2kb_t) \frac{w_0^4}{2\pi} \int_0^{2\pi} d\phi \int_k^{\infty} d\eta \eta^3$$
$$\times \exp\left[-\frac{\eta^2 w_0^2}{2}\left(1 - \frac{b_t}{b}\right)\right](\eta^2 - k^2)^{-1/2} \tag{66}$$

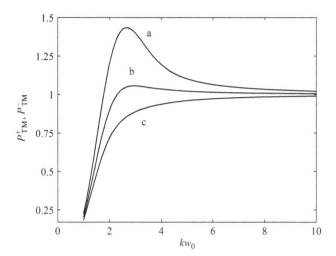

Fig. 2. Time-averaged power $P^{\pm}_{TM}$ transported by the cylindrically symmetric TM full Gaussian wave in the $\pm z$ direction as a function of kw_0 for $1 < kw_0 < 10$, and (a) $b_t/b = 0$, (b) $b_t/b = 0.5$, and (c) $b_t/b = 1.0$. Time-averaged power $P^{\pm}_0$ carried by the corresponding paraxial beam in the $\pm z$ direction is $P^{\pm}_0 = 1$ W.

The ϕ integration is carried out and the variable η is changed as $\eta^2 = k^2(1 + \tau^2)$. Then, Eq. (66) is transformed as

$$P_{im} = -k^4 w_0^4 \exp\left[-\frac{k^2 w_0^2}{2}\left(1 + \frac{b_t}{b}\right)\right]\int_0^\infty d\tau(1 + \tau^2)$$
$$\times \exp\left[-\frac{k^2 w_0^2}{2}\left(1 - \frac{b_t}{b}\right)\tau^2\right] \tag{67}$$

The τ integration is carried out, with the result that

$$P_{im} = -\left(\frac{\pi}{2}\right)^{1/2} k^3 w_0^3 \left(1 - \frac{b_t}{b}\right)^{-1/2} \exp\left[-\frac{k^2 w_0^2}{2}\left(1 + \frac{b_t}{b}\right)\right]$$
$$\times \left[1 + \frac{1}{k^2 w_0^2(1 - b_t/b)}\right] \tag{68}$$

An expression for P_{im} is obtained in terms of b_t/b and kw_0. In general, P_{im} can be positive or negative; here, it is negative. For a fixed kw_0, near $b_t/b = 1$, $|P_{im}|$ increases rapidly and, as anticipated, becomes infinite for $b_t/b = 1$.

The magnitude of the reactive power is shown in Fig. 3 as a function of b_t/b for $0 \le b_t/b \le 1$ and $kw_0 = 1.563$. The real power, $P_{re} = P^+_{TM} + P^-_{TM}$, decreases as b_t/b is increased from 0 to 1 for a fixed kw_0 (see Fig. 1). But $|P_{im}|$ decreases gradually, reaches a minimum, and then increases as b_t/b is increased from 0 to 1. As b_t/b becomes closer to 1, $|P_{im}|$ increases rapidly as b_t/b approaches 1 and becomes infinite for $b_t/b = 1$. In Fig. 4, $|P_{im}|$ is shown as a function of kw_0 for $0.2 < kw_0 < 10$ and $b_t/b = 0.5$. For $b_t/b < 1$, $|P_{im}|$ increases, reaches a maximum, and then decreases and approaches zero as kw_0 is increased.

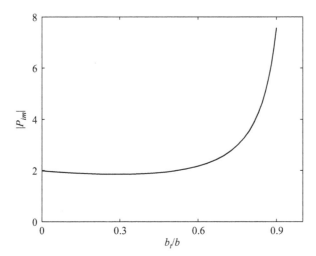

Fig. 3.   Magnitude of the reactive power $|P_{im}|$ of the cylindrically symmetric TM full Gaussian wave as a function of b_t/b for $0 < b_t/b < 1$ and $kw_0 = 1.563$. Time-averaged power in the corresponding paraxial beam is given by $P_0^{\pm} = 1$ W.

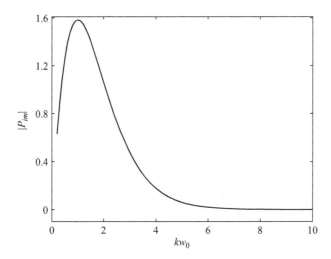

Fig. 4.   Magnitude of the reactive power $|P_{im}|$ of the cylindrically symmetric TM full Gaussian wave as a function of kw_0 for $0.2 < kw_0 < 10$ and $b_t/b = 0.5$. Time-averaged power in the corresponding paraxial beam is given by $P_0^{\pm} = 1$ W.

In the paraxial approximation ($kw_0 \to \infty$), the reactive power $P_{im} = 0$. Figure 4 shows that the paraxial beam result is attained for kw_0 as small as 6.

6 Radiation intensity distribution

From Eqs. (64) and (65), it follows that $\Phi_{\text{TM}}(\theta^+, \phi)$ and $\Phi_{\text{TM}}(\theta^-, \phi)$ are the radiation intensities of the TM Gaussian wave for $z > 0$ and $z < 0$, respectively. Here, θ^+ and θ^- are defined with respect to the $+z$ and $-z$ axes, respectively. In view of Eq. (62), $\mathbf{\Phi}_{TM}(\theta^-, \phi)$ for $z < 0$ is the mirror image of $\mathbf{\Phi}_{TM}(\theta^+, \phi)$ for $z > 0$ about the $z = 0$ plane. Hence, only the radiation intensity distribution for $z > 0$ is discussed.

$\mathbf{\Phi}_{TM}(\theta^+, \phi)$ is independent of ϕ. An important feature is that $\mathbf{\Phi}_{TM}(\theta^+, \phi)$ has a null in the propagation direction. Therefore, w_0 may be considered as the width of the null of the input distribution as given by Eq. (10).

In Figs. 5 and 6, the radiation intensity patterns $\mathbf{\Phi}_{TM}(\theta^+, \phi)$ of the TM Gaussian wave are depicted as functions of θ^+ for $0° < \theta^+ < 90°$ and for the two extreme values of b_t/b, namely $b_t/b = 0$ and $b_t/b = 1$. The radiation intensity pattern $\mathbf{\Phi}_0(\theta^+, \phi)$ of the corresponding paraxial beam is also shown for the purpose of comparison. For Fig. 5, $kw_0 = 1.563$, and for Fig. 6, $kw_0 = 2.980$. $\mathbf{\Phi}_0(\theta^+, \phi)$ is also independent of ϕ. The time-averaged power transported by the corresponding paraxial beam in the $+z$ and $-z$ directions is given by $P_0^\pm = 1$ W. $\mathbf{\Phi}_0(\theta^+, \phi)$ is a good approximation to $\mathbf{\Phi}_{TM}(\theta^+, \phi)$ only in the paraxial region. This result is to be expected because the full wave is constructed so that in the paraxial approximation the full wave reduces to the same paraxial beam for all b_t/b in the range $0 < b_t/b < 1$. If the quality of the paraxial beam approximation to the full wave is judged on the basis of the power transported, as discussed previously, the quality of the paraxial beam approximation of the full wave for a fixed kw_0 improves as b_t/b is decreased from 1 to 0. The radiation intensity patterns of the TM Gaussian wave are essentially the same for all values of b_t/b in the range $0 < b_t/b < 1$ in the paraxial region, but they have significant differences outside the paraxial region, particularly close to the broadside region corresponding to $\theta^+ = 90°$. A comparison of Figs. 5 and 6 reveals that the sharpness of the null of the radiation intensity pattern in the propagation direction increases as kw_0 is increased.

Only the TM Gaussian wave is treated. The TE Gaussian wave differs from the TM Gaussian wave in the polarization; the TM wave is radially polarized, but the TE wave is

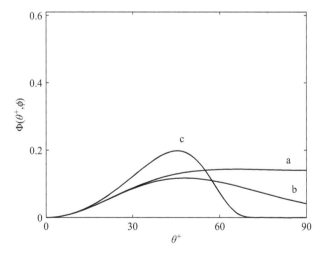

Fig. 5. Radiation intensity pattern $\Phi(\theta^+, \phi)$ for $z > 0$ as a function of θ^+ for $0° < \theta^+ < 90°$ and $kw_0 = 1.563$. (a) TM full Gaussian wave for $b_t/b = 0$, (b) TM full Gaussian wave for $b_t/b = 1$, and (c) corresponding paraxial beam. Time-averaged power in the paraxial beam is given by $P_0^\pm = 1$ W.

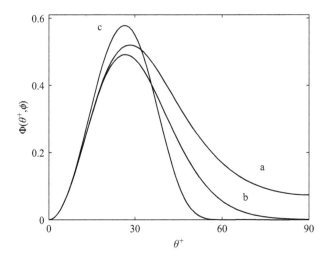

Fig. 6.   Same as Fig. 5, but with $kw_0 = 2.980$.

azimuthally polarized. For the TM wave, the secondary source is an azimuthally directed sheet of magnetic current, but for the TE wave, the source is an azimuthally directed sheet of electric current. The characteristics of the real power, the reactive power, and the radiation intensity distributions all remain unchanged.

For the extended $(0 \leq b_t \leq b)$ full Gaussian wave (Chapter 6), together with the special cases $(b_t = 0)$ of the fundamental Gaussian wave (Chapter 2) and $(b_t = b)$ of the basic full Gaussian wave (Chapter 4), the radiation intensity pattern has a peak in the propagation direction $(\pm z)$. In contrast, for the cylindrically symmetric, TM/TE extended full Gaussian waves, the radiation intensity pattern has a null in the propagation direction. These waves are identified as "hollow" waves; in contrast, the extended full Gaussian waves, which have their radiation intensity peaks in the propagation direction, may be named "solid" waves. The hollow waves are finding applications in practice. The hollow TM/TE waves are cylindrically symmetrical, but the radiation intensity pattern of the solid extended full Gaussian waves varies in the azimuthal direction. The cylindrically symmetric hollow waves do not carry orbital angular momentum. The analysis can be extended to the azimuthally varying modes. The azimuthally varying hollow waves, which carry orbital angular momentum, have applications to the investigation of the guiding and focusing of atoms [9].

References

1. L. W. Davis and G. Patsakos, "TM and TE electromagnetic beams in free space," *Opt. Lett.* **6**, 22–23 (1981).

2. S. R. Seshadri, "Partially coherent Gaussian Schell-model electromagnetic beams," *J. Opt. Soc. Am. A* **16**, 1373–1380 (1999). Section 3.

3. S. R. Seshadri, "Electromagnetic Gaussian beam," *J. Opt. Soc. Am. A* **15**, 2712–2719 (1998). Section 2.

4. S. R. Seshadri, "Quality of paraxial electromagnetic beams," *Appl. Opt.* **45**, 5335–5345 (2006). Appendix A.2.

5. C. J. R. Sheppard and S. Saghafi, "Transverse-electric and transverse-magnetic beam modes beyond the paraxial approximation," *Opt. Lett.* **24**, 1543–1545 (1999).

6. G. A. Deschamps, "Gaussian beam as a bundle of complex rays," *Electron. Lett.* **7**, 684–685 (1971).

7. L. B. Felsen, "Evanescent waves," *J. Opt. Soc. Am.* **66**, 751–760 (1976).

8. W. Magnus and F. Oberhettinger, *Functions of Mathematical Physics* (Chelsea Publishing Company, New York, 1954), p. 35.

9. J. Arlt, T. Hitami, and K. Dholakia, "Atom guiding along Laguerre-Gaussian and Bessel light beams," *Appl. Phys. B, Laser and Optics,* **71**, 549–554 (2000).

CHAPTER 8

Two higher-order full Gaussian waves

Deschamps [1] and Felsen [2] introduced the complex space point source for the full-wave generalization of the fundamental Gaussian beam. Postulation of the required complex space source for other than the fundamental Gaussian beam is difficult. Such a difficult task was achieved recently for two higher-order Gaussian beams. Bandres and Gutierrez-Vega [3] discovered the higher-order point source in the complex space for the full-wave generalization of the higher-order hollow Gaussian beam. Zhang, Song, Chen, Ji, and Shi [4] discovered the system of point sources in the complex space for obtaining the full wave corresponding to the cosh-Gauss paraxial beam [5].

A treatment of the higher-order hollow Gaussian beam is presented. The required higher-order point source in the complex space is introduced. The higher-order hollow full Gaussian wave generated by the complex space source is derived. The basic full Gaussian wave is obtained as a special case of the higher-order hollow full Gaussian wave. The cosh-Gauss paraxial beam is deduced by the two-dimensional Fourier transform technique. The required array of point sources situated on the corners of a square in the complex space is introduced. The full-wave generalization of the cosh-Gauss beam is determined. The basic full Gaussian wave is obtained as a special case of the full cosh-Gauss wave as the side of the square reduces to zero.

1 Higher-order hollow Gaussian wave

1.1 Paraxial beam

The higher-order hollow Gaussian beam varies in the azimuthal direction. In the wave function representing the beam, the azimuthal dependence is separated out as:

$$S^{\pm}(\rho, \phi, z) = \text{trig}(m\phi) S_{0,m}^{\pm}(\rho, z) \tag{1}$$

where $\text{trig}(m\phi)$ stands for $\cos m\phi$ or $\sin m\phi$ and m is a positive integer starting from 0 for $\cos m\phi$ and from 1 for $\sin m\phi$. The mode number in the radial (ρ) and the azimuthal (ϕ)

directions are denoted by 0 and m, respectively. The wave function $S_{0,m}^{\pm}(\rho, z)$ satisfies the reduced Helmholtz equation:

$$\left(\frac{\partial^2}{\partial\rho^2} + \frac{1}{\rho}\frac{\partial}{\partial\rho} - \frac{m^2}{\rho^2} + \frac{\partial^2}{\partial z^2} + k^2\right) S_{0,m}^{\pm}(\rho, z) = 0 \tag{2}$$

Propagation is outward in the $+z$ direction in $z > 0$ and in the $-z$ direction in $z < 0$. The plane-wave phase factor $\exp(\pm ikz)$ is separated out as

$$S_{0,m}^{\pm}(\rho, z) = \exp(\pm ikz)s_{0,m}^{\pm}(\rho, z) \tag{3}$$

In the paraxial approximation, the slowly varying amplitude $S_{0,m}^{\pm}(\rho, z)$ satisfies the paraxial wave equation:

$$\left(\frac{\partial^2}{\partial\rho^2} + \frac{1}{\rho}\frac{\partial}{\partial\rho} - \frac{m^2}{\rho^2} \pm 2ik\frac{\partial}{\partial z}\right) s_{0,m}^{\pm}(\rho, z) = 0 \tag{4}$$

The sign $\pm$, the superscript $\pm$, and the subscript $\pm$ denote that propagation is in the $\pm z$ direction. For the excitation of the higher-order hollow Gaussian beam, the input distribution is assumed as

$$s_{0,m}^{\pm}(\rho, 0) = S_{0,m}^{\pm}(\rho, 0) = 2^{m/2}\frac{\rho^m}{w_0^m}\exp\left(-\frac{\rho^2}{w_0^2}\right) \tag{5}$$

where w_0 is the e-folding distance of the Gaussian part of the input distribution. The factor $2^{m/2}$ is included for later convenience. To obtain the paraxial beam on propagation, we shall use the Bessel transform pair with the mth order Bessel function. The Bessel transform of Eq. (5) is found by the use of Eq. (7.19) as

$$\bar{s}_{0,m}^{\pm}(\eta, 0) = \bar{S}_{0,m}^{\pm}(\eta, 0) = 2^{-m/2}\frac{w_0^2}{2}w_0^m\eta^m\exp\left(-\frac{\eta^2 w_0^2}{4}\right) \tag{6}$$

The Bessel transform representation of $S_{0,m}^{\pm}(\rho, z)$ is substituted in Eq. (4) and the differential equation satisfied by $\bar{s}_{0,m}^{\pm}(\eta, z)$ is found as

$$\left(\frac{\partial}{\partial z} \pm \frac{i\eta^2 w_0^2}{4b}\right)\bar{s}_{0,m}^{\pm}(\eta, z) = 0 \tag{7}$$

where $b = \frac{1}{2}kw_0^2$ is the Rayleigh distance. The solution of Eq. (7) is

$$\bar{s}_{0,m}^{\pm}(\eta, z) = \bar{s}_{0,m}^{\pm}(\eta, 0)\exp\left(-\frac{\eta^2 w_0^2}{4}\frac{i|z|}{b}\right) \tag{8}$$

From Eq. (6), $\bar{s}_{0,m}^{\pm}(\eta, 0)$ is substituted into Eq. (8). The inverse Bessel transform of $\bar{s}_{0,m}^{\pm}(\eta, z)$ is performed to find $S_{0,m}^{\pm}(\rho, z)$ as

$$s_{0,m}^{\pm}(\rho,z) = 2^{-m/2}\frac{w_0^2}{2}w_0^m\int_0^{\infty}d\eta\eta J_m(\eta\rho)\eta^m\exp\left(-\frac{\eta^2 w_0^2}{4q_{\pm}^2}\right) \tag{9}$$

where

$$q_{\pm} = (1 + i|z|/b)^{-1/2} \tag{10}$$

Then, by the use of Eq. (3), $S_{0,m}^{\pm}(\rho,z)$ is expressed as

$$S_{0,m}^{\pm}(\rho,z) = \exp(\pm ikz)2^{-m/2}\frac{w_0^2}{2}w_0^m\int_0^{\infty}d\eta\eta J_m(\eta\rho)\eta^m\exp\left(-\frac{\eta^2 w_0^2}{4q_{\pm}^2}\right) \tag{11}$$

The integral in Eq. (11) is evaluated by the use of Eq. (7.19), and the result is

$$S_{0,m}^{\pm}(\rho,z) = \exp(\pm ikz)2^{m/2}q_{\pm}^{m+2}\left(\frac{q_{\pm}\rho}{w_0}\right)^m\exp\left(-\frac{q_{\pm}^2\rho^2}{w_0^2}\right) \tag{12}$$

The wave function representing the higher-order hollow Gaussian paraxial beam is given by Eq. (1) together with Eq. (12). For the azimuthal mode number $m = 0$, Eq. (5) reduces to the input ($z = 0$) field distribution of the fundamental Gaussian beam, and Eqs. (1) and (12) reproduce the fundamental Gaussian beam for $|z| > 0$.

1.2 Complex space source

Bandres and Gutierrez-Vega [3] discovered the higher-order point source in the complex space that produces the full-wave generalization of the higher-order hollow Gaussian beam. The higher-order point source is situated at $|z| - ib = 0$. This complex space source is given by

$$C_{s,m}(\rho,z) = 2^{-m/2}\frac{w_0^2}{2}w_0^m(-1)^m\rho^m\left(\frac{1}{\rho}\frac{\partial}{\partial\rho}\right)^m\left[\frac{\delta(\rho)}{\rho}\right] \quad \text{for } |z| - ib = 0 \tag{13}$$

For $m = 0$, Eq. (13) reduces to

$$C_{s,0}(\rho,z) = \pi w_0^2\frac{\delta(\rho)}{2\pi\rho} \quad \text{for } m = 0 \text{ and } |z| - ib = 0 \tag{14}$$

The complex space source given by Eq. (13) reduces correctly to that for the fundamental Gaussian beam as given by Eq. (5.5), except for the normalization factor N/ik, which is not included here for the higher-order hollow Gaussian beam. A crucial step is to express Eq. (13) as an inverse Bessel transform with mth order Bessel function. By using Eq. (7.38), Eq. (13) is stated as an inverse Bessel transform with zero order Bessel function as

$$C_{s,m}(\rho,z) = 2^{-m/2}\frac{w_0^2}{2}w_0^m(-1)^m\rho^m\left(\frac{1}{\rho}\frac{\partial}{\partial\rho}\right)^m\int_0^{\infty}d\eta\eta J_0(\eta\rho) \quad \text{for } |z| - ib = 0 \tag{15}$$

In order to transform Eq. (15) as an inverse Bessel transform with mth order Bessel function, let us consider the operator $(1/\rho)(\partial/\partial\rho)$ acting on $J_m(\eta\rho)/\rho^m$. The use of the Bessel transform relation [6]

$$\frac{\partial J_m(\eta\rho)}{\partial\rho} - \frac{m}{\rho}J_m(\eta\rho) = -\eta J_{m+1}(\eta\rho) \tag{16}$$

yields the result of the stated operation as follows:

$$\frac{1}{\rho}\frac{\partial}{\partial\rho}\left[\frac{J_m(\eta\rho)}{\rho^m}\right] = -\eta\frac{J_{m+1}(\eta\rho)}{\rho^{m+1}} \tag{17}$$

Starting from $m = 0$ and repeating the operation m times, it is found from Eq. (17) that

$$\left(\frac{1}{\rho}\frac{\partial}{\partial\rho}\right)^m[J_0(\eta\rho)] = (-1)^m\eta^m\frac{J_m(\eta\rho)}{\rho^m} \tag{18}$$

The substitution of Eq. (18) into Eq. (15) yields

$$C_{s,m}(\rho,z) = 2^{-m/2}\frac{w_0^2}{2}w_0^m\int_0^\infty d\eta\eta J_m(\eta\rho)\eta^m \tag{19}$$

The source given by Eq. (19) located at $|z| - ib = 0$ produces the paraxial beam for $|z| > 0$. The same source moved to $|z| = 0$ generates the asymptotic value of the paraxial beam. The paraxial wave equation is only an approximation for the full Helmholtz wave equation. Therefore, the complex space source deduced from the paraxial beam, except for an excitation coefficient S_{ex}, is used for the Helmholtz equation to obtain the asymptotic value of the full Gaussian wave. Hence, the complex space source for obtaining the full-wave generalization of the higher-order hollow Gaussian beam is found from Eq. (19) as

$$C_{s,m}(\rho,z) = S_{ex}2^{-m/2}\frac{w_0^2}{2}w_0^m\int_0^\infty d\eta\eta J_m(\eta\rho)\eta^m \tag{20}$$

S_{ex} is determined by the requirement in the paraxial approximation that the full wave should reproduce correctly the initially chosen paraxial beam.

The source that produces the asymptotic value of the full-wave solution of the Helmholtz equation is obtained from Eq. (20) as

$$C_{s\infty,m}(\rho,z) = S_{ex}2^{-m/2}\frac{w_0^2}{2}w_0^m\delta(z)\int_0^\infty d\eta\eta J_m(\eta\rho)\eta^m \tag{21}$$

1.3 Hollow Gaussian wave

Let $G_{0,m}(\rho,z)$ be the solution of the reduced Helmholtz equation [Eq. (2)] for the source given by Eq. (21). Then, $G_{0,m}(\rho,z)$ satisfies the following differential equation:

$$\left(\frac{\partial^2}{\partial\rho^2} + \frac{1}{\rho}\frac{\partial}{\partial\rho} - \frac{m^2}{\rho^2} + \frac{\partial^2}{\partial z^2} + k^2\right)G_{0,m}(\rho,z) = -S_{ex}2^{-m/2}\frac{w_0^2}{2}w_0^m\delta(z)\int_0^\infty d\eta\eta J_m(\eta\rho)\eta^m \tag{22}$$

This differential equation is solved in the same way as Eq. (7.44). The difference is that the Bessel transform with $J_m(\eta\rho)$ is used instead of $J_1(\eta\rho)$. The result is

$$G_{0,\mathrm{m}}(\rho,z) = \frac{iS_{ex}}{2} 2^{-m/2} \frac{w_0^2}{2} w_0^m \int_0^\infty d\eta\eta J_m(\eta\rho)\eta^m \zeta^{-1} \exp(i\zeta|z|) \tag{23}$$

where

$$\zeta = (k^2 - \eta^2)^{1/2} \tag{24}$$

$G_{0,\mathrm{m}}(\rho,z)$ is analytically continued from $|z|$ to $|z| - ib$ to obtain $S^\pm_{0,\mathrm{m}}(\rho,z)$ as

$$S^\pm_{0,\mathrm{m}}(\rho,z) = \frac{iS_{ex}}{2} 2^{-m/2} \frac{w_0^2}{2} w_0^m \int_0^\infty d\eta\eta J_m(\eta\rho)\eta^m \zeta^{-1} \exp[i\zeta(|z| - ib)] \tag{25}$$

The paraxial approximation corresponds to $\eta^2 \ll k^2$. The paraxial approximation of Eq. (25) is determined as

$$S^\pm_{0,\mathrm{m}}(\rho,z) = \exp(\pm ikz) \frac{iS_{ex}}{2k} \exp(kb) 2^{-m/2} \frac{w_0^2}{2} w_0^m \int_0^\infty d\eta\eta J_m(\eta\rho)\eta^m \exp\left(-\frac{\eta^2 w_0^2}{4q_\pm^2}\right) \tag{26}$$

The excitation coefficient is chosen as

$$S_{ex} = -i2k\exp(-kb) \tag{27}$$

Then, Eq. (26) is the same as Eq. (11), and the paraxial approximation of Eq. (25) correctly reproduces the initially chosen paraxial beam. Substituting S_{ex} from Eq. (27) into Eq. (25) yields the full-wave generalization of the higher-order hollow Gaussian beam given by Eq. (11) as

$$S^\pm_{0,\mathrm{m}}(\rho,z) = k\exp(-kb) 2^{-m/2} \frac{w_0^2}{2} w_0^m \int_0^\infty d\eta\eta J_m(\eta\rho)\eta^m \zeta^{-1} \exp[i\zeta(|z| - ib)] \tag{28}$$

The full-wave generalization expressed by Eqs. (1) and (28) is designated the higher-order hollow full Gaussian wave.

For $m = 0$, by using Eqs. (A15), (A16), and (A36), Eq. (28) is simplified to yield

$$S^\pm_{0,0}(\rho,z) = \left[-\pi w_0^2 2ik\exp(-kb)\right] \frac{\exp\left\{ik\left[\rho^2 + (|z| - ib)^2\right]^{1/2}\right\}}{4\pi\left[\rho^2 + (|z| - ib)^2\right]^{1/2}} \tag{29}$$

The higher-order hollow full Gaussian wave as given by Eq. (28) reduces correctly to the basic full Gaussian wave as given by Eq. (4.5), except for the normalization factor N/ik, which is not included for the higher-order full Gaussian wave. If the normalization factor N/ik is included, for $m = 0$, Eq. (28) yields the fundamental Gaussian beam in the paraxial approximation and the basic full Gaussian wave in the full-wave generalization.

2 cosh-Gauss wave

2.1 cosh-Gauss beam

Casperson, Hall, and Tovar [5] introduced the cosh-Gauss beam. For generating the linearly polarized cosh-Gauss beam, the paraxial approximation of the x component of the magnetic vector potential on the input plane $(z = 0)$ is assumed as

$$A_{x0}^{\pm}(x, y, 0) = \frac{N}{ik}\exp\left(-\frac{x^2 + y^2}{w_0^2}\right)\exp\left(-\frac{a^2}{2w_0^2}\right)\cosh\left(\frac{ax}{w_0^2}\right)\cosh\left(\frac{ay}{w_0^2}\right) \tag{30}$$

where N is a normalization constant, w_0 is the $e-$ folding distance of the Gaussian part of the distribution, and a is the length parameter. The factor $\exp(-a^2/2w_0^2)$ is included for convenience. The cosh terms are expanded and Eq. (30) is rearranged to obtain

$$A_{x0}^{\pm}(x, y, 0) = \frac{N}{ik}\frac{1}{4}\left(\exp\left\{-\frac{1}{w_0^2}\left[\left(x - \frac{a}{2}\right)^2 + \left(y - \frac{a}{2}\right)^2\right]\right\}\right.$$

$$+ \exp\left\{-\frac{1}{w_0^2}\left[\left(x - \frac{a}{2}\right)^2 + \left(y + \frac{a}{2}\right)^2\right]\right\}$$

$$+ \exp\left\{-\frac{1}{w_0^2}\left[\left(x + \frac{a}{2}\right)^2 + \left(y - \frac{a}{2}\right)^2\right]\right\}$$

$$+ \left.\exp\left\{-\frac{1}{w_0^2}\left[\left(x + \frac{a}{2}\right)^2 + \left(y + \frac{a}{2}\right)^2\right]\right\}\right) \tag{31}$$

The Fourier transform of a typical term in Eq. (31) is evaluated as follows:

$$I_x = \int_{-\infty}^{\infty} dx\exp(i2\pi p_x x)\exp\left[-\frac{1}{w_0^2}\left(x - \frac{a}{2}\right)^2\right]$$

$$= \exp(i\pi p_x a)\int_{-\infty}^{\infty} d\xi\exp(i2\pi p_x \xi)\exp\left(-\frac{\xi^2}{w_0^2}\right) \tag{32}$$

where the variable is changed in accordance with $x = a/2 + \xi$. By using Eqs. (B1) and (B6), the integral in Eq. (32) is evaluated, with the result that

$$I_x = \exp(i\pi p_x a)\pi^{1/2}w_0\exp\left(-\pi^2 w_0^2 p_x^2\right) \tag{33}$$

The other terms in Eq. (31) are treated similarly. The Fourier transform of Eq. (31) is obtained as

$$\overline{A}_{x0}^{\pm}(p_x, p_y, 0) = \frac{N}{ik}\frac{1}{4}\pi w_0^2\exp\left[-\pi^2 w_0^2\left(p_x^2 + p_y^2\right)\right]$$

$$\times \left\{\exp[i\pi(p_x + p_y)a] + \exp[i\pi(p_x - p_y)a]\right.$$

$$+ \left.\exp[-i\pi(p_x - p_y)a] + \exp[-i\pi(p_x + p_y)a]\right\} \tag{34}$$

$A_{x0}^{\pm}(x,y,z)$ is separated out as

$$A_{x0}^{\pm}(x,y,z) = \exp(\pm ikz)a_{x0}^{\pm}(x,y,z) \qquad (35)$$

and the slowly varying amplitude $a_{x0}^{\pm}(x,y,z)$ satisfies the paraxial wave equation:

$$\left(\frac{\partial^2}{\partial x^2} + \frac{\partial^2}{\partial y^2} \pm 2ik\frac{\partial}{\partial z}\right)a_{x0}^{\pm}(x,y,z) = 0 \qquad (36)$$

The Fourier transform representation of $a_{x0}^{\pm}(x,y,z)$ is substituted into Eq. (36) and the following differential equation is obtained:

$$\left[\frac{\partial}{\partial z} \pm \frac{i\pi^2 w_0^2}{b}\left(p_x^2 + p_y^2\right)\right]\overline{a}_{x0}^{\pm}(p_x,p_y,z) = 0 \qquad (37)$$

for the Fourier transform $\overline{a}_{x0}^{\pm}(p_x,p_y,z)$ of $a_{x0}^{\pm}(x,y,z)$. The solution of Eq. (37) is

$$\overline{a}_{x0}^{\pm}(p_x,p_y,z) = \overline{a}_{x0}^{\pm}(p_x,p_y,0)\exp\left[-\pi^2 w_0^2\left(p_x^2 + p_y^2\right)\frac{i|z|}{b}\right] \qquad (38)$$

Equation (35) shows that $\overline{A}_{x0}^{\pm}(p_x,p_y,0) = \overline{a}_{x0}^{\pm}(p_x,p_y,0)$. From Eq. (34), $\overline{a}_{x0}^{\pm}(p_x,p_y,0)$ is substituted into Eq. (38) to find $\overline{a}_{x0}^{\pm}(p_x,p_y,z)$. The inverse Fourier transform is carried out to obtain

$$a_{x0}^{\pm}(x,y,z) = \frac{N}{ik}\frac{\pi w_0^2}{4}\int_{-\infty}^{\infty}\int_{-\infty}^{\infty} dp_x dp_y \exp\left[-\frac{\pi^2 w_0^2\left(p_x^2 + p_x^2\right)}{q_{\pm}^2}\right]$$

$$\times \left\{\exp\left[-i2\pi p_x\left(x - \frac{a}{2}\right) - i2\pi p_y\left(y - \frac{a}{2}\right)\right]\right.$$

$$+ \exp\left[-i2\pi p_x\left(x - \frac{a}{2}\right) - i2\pi p_y\left(y + \frac{a}{2}\right)\right]$$

$$+ \exp\left[-i2\pi p_x\left(x + \frac{a}{2}\right) - i2\pi p_y\left(y - \frac{a}{2}\right)\right]$$

$$\left.+ \exp\left[-i2\pi p_x\left(x + \frac{a}{2}\right) - i2\pi p_y\left(y + \frac{a}{2}\right)\right]\right\} \qquad (39)$$

The integrals are evaluated by the use of Eqs. (B1) and (B6) to find $a_{x0}^{\pm}(x,y,z)$. Then, from Eqs. (35) and (39), $A_{x0}^{\pm}(x,y,z)$ is found as

$$A_{x0}^{\pm}(x,y,z) = \exp(\pm ikz)\frac{N}{ik}\frac{1}{4}q_{\pm}^2$$

$$\times \left(\exp\left\{-\frac{q_{\pm}^2}{w_0^2}\left[\left(x-\frac{a}{2}\right)^2 + \left(y-\frac{a}{2}\right)^2\right]\right\}\right.$$

$$+ \exp\left\{-\frac{q_{\pm}^2}{w_0^2}\left[\left(x-\frac{a}{2}\right)^2 + \left(y+\frac{a}{2}\right)^2\right]\right\}$$

$$+ \exp\left\{-\frac{q_{\pm}^2}{w_0^2}\left[\left(x+\frac{a}{2}\right)^2 + \left(y-\frac{a}{2}\right)^2\right]\right\}$$

$$\left.+ \exp\left\{-\frac{q_{\pm}^2}{w_0^2}\left[\left(x+\frac{a}{2}\right)^2 + \left(y+\frac{a}{2}\right)^2\right]\right\}\right) \tag{40}$$

The vector potential representing the cosh-Gauss beam is given by Eq. (40). For $a = 0$, Eq. (30) reduces to the input ($z = 0$) field distribution of the fundamental Gaussian beam; similarly, for $a = 0$, Eq. (40) reproduces the fundamental Gaussian beam for $|z| > 0$.

2.2 Complex space source

Zhang, Song, Chen, Ji, and Shi [4] discovered an array of point sources in the complex space that generates the basic full-wave generalization of the cosh-Gauss beam. More recently, Sheppard has obtained similar results for the cosine-Gauss beam [7]. The point sources are situated at $|z| - ib = 0$. There are four point sources situated at the corners of a square of side a, with the center at the origin; the sides of the square are parallel to the x and y axes. All the point sources are of identical strength: $(N/ik)(\pi w_0^2/4)$. This system of complex space point sources is given by

$$C_s(x,y,z) = \frac{N}{ik}\frac{\pi w_0^2}{4}\left[\delta\left(x-\frac{a}{2}\right)\delta\left(y-\frac{a}{2}\right)\right.$$

$$+ \delta\left(x-\frac{a}{2}\right)\delta\left(y+\frac{a}{2}\right) + \delta\left(x+\frac{a}{2}\right)\delta\left(y-\frac{a}{2}\right)$$

$$\left.+ \delta\left(x+\frac{a}{2}\right)\delta\left(y+\frac{a}{2}\right)\right] \quad \text{for } |z| - ib = 0 \tag{41}$$

For $a = 0$, Eq. (41) reduces to

$$C_s(x,y,z) = \frac{N}{ik}\pi w_0^2\delta(x)\delta(y) \quad \text{for } a = 0 \text{ and } |z| - ib = 0 \tag{42}$$

The system of complex space point sources given by Eq. (41) reduces correctly to that for the fundamental Gaussian beam as given by Eq. (5.5). The source given by Eq. (41) situated at $|z| - ib = 0$ produces the paraxial beam for $|z| > 0$. The same source at $|z| = 0$ generates the asymptotic field of the paraxial beam. The paraxial wave equation is only an approximation for the full Helmholtz wave equation. Therefore, the complex space source derived from the paraxial beam, except for an excitation coefficient S_{ex}, is used for the Helmholtz equation to deduce the asymptotic value of the cosh-Gauss full wave. Hence, the complex space source for obtaining the full-wave generalization of the cosh-Gauss beam is obtained from Eq. (41) as

$$C_s(x, y, z) = S_{ex} \frac{N}{ik} \frac{\pi w_0^2}{4} \left[\delta\left(x - \frac{a}{2}\right)\delta\left(y - \frac{a}{2}\right) \right.$$

$$+ \delta\left(x - \frac{a}{2}\right)\delta\left(y + \frac{a}{2}\right) + \delta\left(x + \frac{a}{2}\right)\delta\left(y - \frac{a}{2}\right)$$

$$\left. + \delta\left(x + \frac{a}{2}\right)\delta\left(y + \frac{a}{2}\right) \right] \quad \text{for } |z| - ib = 0 \quad (43)$$

The requirement in the paraxial approximation that the full wave should correctly reproduce the initially chosen paraxial beam enables us to determine S_{ex}.

The source that produces the asymptotic value of the full-wave solution of the Helmholtz equation is obtained from Eq. (43) as

$$C_s(x, y, z) = S_{ex} \frac{N}{ik} \frac{\pi w_0^2}{4} \delta(z) \left[\delta\left(x - \frac{a}{2}\right)\delta\left(y - \frac{a}{2}\right) \right.$$

$$+ \delta\left(x - \frac{a}{2}\right)\delta\left(y + \frac{a}{2}\right) + \delta\left(x + \frac{a}{2}\right)\delta\left(y - \frac{a}{2}\right)$$

$$\left. + \delta\left(x + \frac{a}{2}\right)\delta\left(y + \frac{a}{2}\right) \right] \quad (44)$$

2.3 cosh-Gauss wave

Let $G(x, y, z)$ be the solution of the Helmholtz equation for the source given by Eq. (44). The source consists of four-point sources of strength $S_{ex}(N/ik)(\pi w_0^2/4)$. The point sources are located at $x = \pm a/2$, $y = \pm a/2$, and $z = 0$. In Eq. (A1), the corresponding equation is given for a point source of unit strength located at $x = 0$, $y = 0$, and $z = 0$. From Eqs. (A15) and (A16), $G(x, y, z)$ corresponding to the source given by Eq. (44) is obtained. $G(x, y, z)$ is analytically continued from $|z|$ to $|z| - ib$ to find the exact vector potential as

$$A_x^{\pm}(x, y, z) = S_{ex} \frac{N}{ik} \frac{\pi w_0^2}{4} \left[\frac{\exp(ikr_{++})}{4\pi r_{++}} + \frac{\exp(ikr_{+-})}{4\pi r_{+-}} + \frac{\exp(ikr_{-+})}{4\pi r_{-+}} + \frac{\exp(ikr_{--})}{4\pi r_{--}} \right] \quad (45)$$

where

$$r_{++} = \left[\left(x - \frac{a}{2}\right)^2 + \left(y - \frac{a}{2}\right)^2 + (|z| - ib)^2 \right]^{1/2} \quad (46)$$

$$r_{+-} = \left[\left(x - \frac{a}{2}\right)^2 + \left(y + \frac{a}{2}\right)^2 + (|z| - ib)^2 \right]^{1/2} \quad (47)$$

$$r_{-+} = \left[\left(x + \frac{a}{2} \right)^2 + \left(y - \frac{a}{2} \right)^2 + (|z| - ib)^2 \right]^{1/2} \tag{48}$$

$$r_{--} = \left[\left(x + \frac{a}{2} \right)^2 + \left(y + \frac{a}{2} \right)^2 + (|z| - ib)^2 \right]^{1/2} \tag{49}$$

The paraxial approximation corresponds to $(x \pm a/2)^2 + (y \pm a/2)^2 \ll |(|z| - ib)^2|$. The paraxial approximation of Eq. (45) is determined as

$$A_{x0}^{\pm}(x, y, z) = \exp(\pm ikz) S_{ex} \frac{N}{ik} \frac{\pi w_0^2}{4} \frac{\exp(kb)}{(-4\pi ib)} q_{\pm}^2$$

$$\times \left(\exp\left\{ -\frac{q_{\pm}^2}{w_0^2} \left[\left(x - \frac{a}{2} \right)^2 + \left(y - \frac{a}{2} \right)^2 \right] \right\} + \text{three similar terms} \right) \tag{50}$$

The excitation coefficient is chosen to be the same as in Eq. (27). Then, Eq. (50) is the same as Eq. (40). Therefore, the paraxial approximation of Eq. (45) correctly reproduces the initially chosen paraxial beam. Substituting S_{ex} from Eq. (27) into Eq. (45) yields the full-wave generalization of the cosh-Gauss beam as

$$A_x^{\pm}(x, y, z) = \frac{N}{ik} \frac{[-\pi w_0^2 2ik \exp(-kb)]}{4} \left[\frac{\exp(ikr_{++})}{4\pi r_{++}} \right.$$

$$\left. + \frac{\exp(ikr_{+-})}{4\pi r_{+-}} + \frac{\exp(ikr_{-+})}{4\pi r_{-+}} + \frac{\exp(ikr_{--})}{4\pi r_{--}} \right] \tag{51}$$

The full-wave generalization of the cosh-Gauss paraxial beam as given by Eq. (51) is designated the basic full cosh-Gauss wave.

For $a = 0$, Eq. (51) reduces to

$$A_x^{\pm}(x, y, z) = \frac{N}{ik} [-\pi w_0^2 2ik \exp(-kb)] \frac{\exp\left\{ ik \left[x^2 + y^2 + (|z| - ib)^2 \right]^{1/2} \right\}}{4\pi \left[x^2 + y^2 + (|z| - ib)^2 \right]^{1/2}} \tag{52}$$

For $a = 0$, Eq. (30) becomes the input field distribution of the fundamental Gaussian beam [see Eq. (1.1)]. For $a = 0$, Eq. (40) reproduces the fundamental Gaussian beam in the paraxial approximation [see Eq. (1.11)], and Eq. (51) reduces to the basic full Gaussian wave in the full-wave generalization [see Eq. (4.5)].

References

1. G. A. Deschamps, "Gaussian beam as a bundle of complex rays," *Electron. Lett.* **7**, 684–685 (1971).

2. L. B. Felsen, "Evanescent waves," *J. Opt. Soc. Am.* **66**, 751–760 (1976).

3. M. A. Bandres and J. C. Gutierrez-Vega, "Higher-order complex source for elegant Laguerre-Gaussian waves," *Opt. Lett.* **29**, 2213–2215 (2004).

4. Y. Zhang, Y. Song, Z. Chen, J. Ji, and Z. Shi, "Virtual sources for cosh-Gaussian beam," *Opt. Lett.* **32**, 292–294 (2007).

5. L. W. Casperson, D. G. Hall, and A. A. Tovar , "Sinusoidal-Gaussian beams in complex optical systems," *J. Opt. Soc. Am. A* **14**, 3341–3348 (1997).

6. W. Magnus and F. Oberhettinger, *Functions of Mathematical Physics* (Chelsea Publishing Company, New York, 1954), p. 16.

7. C. J. R. Sheppard, "Complex source point theory of paraxial and nonparaxial cosine-Gauss and Bessel-Gauss beams," *Opt. Lett.* **38**, 564–566 (2013).

Basic full complex-argument Laguerre–Gauss wave

The paraxial wave equation in the cylindrical coordinate system has a series of higher-order solutions known as the complex-argument Laguerre–Gauss beams [1,2]. This series of eigenfunctions is a complete set. These higher-order Gaussian beams are characterized by the radial mode number n and the azimuthal mode number m. The fundamental Gaussian beam is the lowest-order $(n = 0,\ m = 0)$ mode in this set. The higher-order hollow Gaussian beams discussed in Chapter 8 correspond to $n = 0$. In this chapter, a treatment of the complex-argument Laguerre–Gauss beams is presented. As for the fundamental Gaussian beam, for the complex-argument Laguerre-Gauss beams also, the reactive power vanishes. The higher-order point source in the complex space required for the full-wave generalization of the complex-argument Laguerre–Gauss beams is deduced. The basic full complex-argument Laguerre–Gauss wave generated by the complex space source is determined [3,4]. The real and reactive powers of the basic full higher-order wave are evaluated. As expected, the reactive power is infinite, as for the basic full Gaussian wave. The general characteristics of the real power are examined. The real power increases monotonically, approaching the limiting value of the paraxial beam as kw_0 is increased. For a fixed kw_0, in general, the real power decreases as the mode order increases.

1 Complex-argument Laguerre–Gauss beam

1.1 Paraxial beam

The secondary source is an infinitesimally thin sheet of electric current located on the plane $z = 0$. The source generates waves that propagate outward in the $+z$ direction in $z > 0$ and in the $-z$ direction in $z < 0$. The source electric current density is in the x direction and the x component of the magnetic vector potential is excited. The vector potential is continuous across the plane $z = 0$. For the excitation of a linearly polarized electromagnetic complex-argument Laguerre–Gauss beam, the required x component of the magnetic vector potential at the secondary source plane is specified as

$$a_{x0}^{\pm}(\rho, \phi, 0) = A_{x0}^{\pm}(\rho, \phi, 0) = \frac{N_{nm}}{ik}(-1)^n n! 2^{2n+m/2}$$

$$\times \cos m\phi \left(\frac{\rho^2}{w_0^2}\right)^{m/2} L_n^m\left(\frac{\rho^2}{w_0^2}\right) \exp\left(-\frac{\rho^2}{w_0^2}\right) \tag{1}$$

where N_{nm} is the normalization constant, k is the wavenumber, n is the radial mode number, and m is the azimuthal mode number. As indicated, it is possible for the normalization constant to depend on the mode numbers n and m. $L_n^m(\ )$ is the associated or generalized Laguerre polynomial, where n is the order and m is the degree of $L_n^m(\ )$. For $m = 0$, the Laguerre polynomial is denoted by $L_n(\ )$. $L_n^m(x)$ is defined as follows {see formula 8.970.1 in [5]}

$$L_n^m(x) = \frac{1}{n!}x^{-m}e^x\frac{\partial^n}{\partial x^n}[x^{n+m}e^{-x}] \tag{2}$$

$$= \sum_{r=0}^{r=n}\frac{(-1)^r(n+m)!x^r}{(n-r)!r!(m+r)!} \tag{3}$$

For $n = 0$, it is found from Eq. (2) that

$$L_0^m(x) = 1 \tag{4}$$

The medium in $|z| > 0$ is cylindrically symmetrical about the z axis. Therefore, the ϕ dependence of $a_{x0}^{\pm}(\rho, \phi, z)$ and $A_{x0}^{\pm}(\rho, \phi, z)$ is the same $\cos m\phi$ as at the input plane $z = 0$; that is, the ϕ dependence remains unchanged on propagation. For definiteness $\cos m\phi$ is used. The mode number m is a positive integer starting with 0. Also, $\sin m\phi$ may be used instead of $\cos m\phi$; then m starts with 1.

The paraxial approximation of the x component of the magnetic vector potential is given by $A_{x0}^{\pm}(\rho, \phi, z)$. The additional subscript 0 is used to denote paraxial. The rapidly varying phase is separated out to obtain:

$$A_{x0}^{\pm}(\rho, \phi, z) = \exp(\pm ikz)a_{x0}^{\pm}(\rho, \phi, z) \tag{5}$$

The reduced paraxial wave equation satisfied by the slowly varying amplitude $a_{x0}^{\pm}(\rho, \phi, z)$ is given by

$$\left(\frac{\partial^2}{\partial\rho^2} + \frac{1}{\rho}\frac{\partial}{\partial\rho} - \frac{m^2}{\rho^2} \pm 2ik\frac{\partial}{\partial z}\right)a_{x0}^{\pm}(\rho, \phi, z) = 0 \tag{6}$$

As before, the sign $\pm$ and the superscript $\pm$ are used to indicate that the propagation is in the $\pm z$ direction. The mth order Bessel transform representation of $a_{x0}^{\pm}(\rho, \phi, z)$ is substituted into Eq. (6) to obtain

$$\left(\frac{\partial}{\partial z} \pm \frac{\eta^2 w_0^2}{4}\frac{i}{b}\right)\bar{a}_{x0}^{\pm}(\eta, \phi, z) = 0 \tag{7}$$

where $b = \frac{1}{2}kw_0^2$, and $\bar{a}_{x0}^{\pm}(\eta, \phi, z)$ is the mth order Bessel transform of $a_{x0}^{\pm}(\rho, \phi, z)$. The order of the Bessel function is m in the mth order Bessel transform. The solution of Eq. (7) is

$$\bar{a}_{x0}^{\pm}(\eta, \phi, z) = \bar{a}_{x0}^{\pm}(\eta, \phi, 0)\exp\left(-\frac{\eta^2 w_0^2}{4}\frac{i|z|}{b}\right) \tag{8}$$

From Eq. (1), $\bar{a}_{x0}^{\pm}(\eta, \phi, 0)$ is expressed as

$$\bar{a}_{x0}^{\pm}(\eta, \phi, 0) = \frac{N_{nm}}{ik}(-1)^n n! 2^{2n+m/2} \cos m\phi$$
$$\times \int_0^{\infty} d\rho \rho J_m(\eta\rho) \left(\frac{\rho^2}{w_0^2}\right)^{m/2} L_n^m\left(\frac{\rho^2}{w_0^2}\right) \exp\left(-\frac{\rho^2}{w_0^2}\right) \qquad (9)$$

Bandres and Gutierrez-Vega [4] have established the following mth order Bessel transform relations:

$$\int_0^{\infty} d\eta \eta J_m(\eta\rho) \eta^{2n+m} \exp\left(-p^2\eta^2\right) = \frac{n!}{2} \frac{1}{p^{2(n+1+m/2)}} \left(\frac{\rho^2}{4p^2}\right)^{m/2} L_n^m\left(\frac{\rho^2}{4p^2}\right) \exp\left(-\frac{\rho^2}{4p^2}\right) \qquad (10)$$

$$\int_0^{\infty} d\rho \rho J_m(\eta\rho) \left(\frac{\rho^2}{4p^2}\right)^{m/2} L_n^m\left(\frac{\rho^2}{4p^2}\right) \exp\left(-\frac{\rho^2}{4p^2}\right) = \frac{2}{n!} p^{2(n+1+m/2)} \eta^{2n+m} \exp\left(-p^2\eta^2\right) \qquad (11)$$

The corresponding zero order Bessel transform relations are obtained from formula 7.421.2 in [5]. The integral in Eq. (9) is found by the use of Eq. (11), with the result that

$$\bar{a}_{x0}^{\pm}(\eta, \phi, 0) = \frac{N_{nm}}{ik}(-1)^n 2^{-1-m/2} w_0^2 \cos m\phi \left(\eta^2 w_0^2\right)^{n+m/2} \exp\left(-\frac{\eta^2 w_0^2}{4}\right) \qquad (12)$$

From Eq. (12), $\bar{a}_{x0}^{\pm}(\eta, \phi, 0)$ is substituted into Eq. (8). The inverse Bessel transform of the resulting $\bar{a}_{x0}^{\pm}(\eta, \phi, z)$ is expressed as

$$\bar{a}_{x0}^{\pm}(\rho, \phi, z) = \frac{N_{nm}}{ik}(-1)^n 2^{-m/2} \frac{w_0^2}{2} w_0^{2n+m} \cos m\phi \int_0^{\infty} d\eta \eta J_m(\eta\rho) \eta^{2n+m} \exp\left(-\frac{\eta^2 w_0^2}{4q_{\pm}^2}\right) \qquad (13)$$

where

$$q_{\pm} = \left(1 + i\frac{|z|}{b}\right)^{-1/2} \qquad (14)$$

The subscript $\pm$ denotes propagation in the $\pm z$ direction. The integral in Eq. (13) is determined by the use of Eq. (10). The result is

$$a_{x0}^{\pm}(\rho, \phi, z) = \frac{N_{nm}}{ik}(-1)^n n! 2^{2n+m/2} q_{\pm}^{2n+m+2} \cos m\phi \left(\frac{q_{\pm}^2 \rho^2}{w_0^2}\right)^{m/2} L_n^m\left(\frac{q_{\pm}^2 \rho^2}{w_0^2}\right) \exp\left(-\frac{q_{\pm}^2 \rho^2}{w_0^2}\right) \qquad (15)$$

From Eqs. (5) and (15), the complex-argument Laguerre–Gauss beam launched in $|z| > 0$ is determined as

$$A_{x0}^{\pm}(\rho, \phi, z) = \exp\left(\pm ikz\right) \frac{N_{nm}}{ik}(-1)^n n! 2^{2n+m/2} q_{\pm}^{2n+m+2}$$
$$\times \cos m\phi \left(\frac{q_{\pm}^2 \rho^2}{w_0^2}\right)^{m/2} L_n^m\left(\frac{q_{\pm}^2 \rho^2}{w_0^2}\right) \exp\left(-\frac{q_{\pm}^2 \rho^2}{w_0^2}\right) \qquad (16)$$

An important feature of the complex-argument Laguerre–Gauss beam is that it has a closed-form Bessel transform representation, as is seen from Eq. (13).

1.2 Time-averaged power

The electromagnetic fields associated with the paraxial beam are found from Eq. (1.12) as

$$E_{x0}^{\pm}(\rho,\phi,z) = \pm H_{y0}^{\pm}(\rho,\phi,z) = ikA_{x0}^{\pm}(\rho,\phi,z) \tag{17}$$

Equations (16) and (17) show that $A_{x0}^{\pm}(\rho,\phi,z)$ and $E_{x0}^{\pm}(\rho,\phi,z)$ are continuous, but $H_{y0}^{\pm}(\rho,\phi,z)$ is discontinuous across the secondary source plane $z = 0$. The discontinuity of $H_{y0}^{\pm}(\rho,\phi,z)$ is equivalent to an electric current density induced on the $z = 0$ plane. By the use of Eq. (17), the induced electric current density is found as

$$\mathbf{J}_0(\rho,\phi,z) = \hat{z} \times \hat{y}[H_y^+(\rho,\phi,0) - H_y^-(\rho,\phi,0)]\delta(z) = -\hat{x}2ikA_{x0}^{\pm}(\rho,\phi,0)\delta(z) \tag{18}$$

The complex power is obtained by the use of Eqs. (5) and (17) as

$$P_C = P_{re} + iP_{im} = -\frac{c}{2}\int_0^\infty d\rho\rho\int_0^{2\pi} d\phi\int_{-\infty}^\infty dz\mathbf{E}_0^{\pm}(\rho,\phi,z)\cdot\mathbf{J}_0^*(\rho,\phi,z)$$

$$= ck^2\int_0^\infty d\rho\rho\int_0^{2\pi} d\phi a_{x0}^{\pm}(\rho,\phi,0)a_{x0}^{\pm*}(\rho,\phi,0) \tag{19}$$

P_C given by Eq. (19) is real; therefore, P_{im}, the imaginary part of P_C, is zero; that is,

$$P_{im} = 0 \tag{20}$$

The reactive power of the complex-argument Laguerre–Gauss beam vanishes.

The mth order Bessel transform representation is substituted for $a_{x0}^{\pm}(\rho,\phi,0)$ in Eq. (19), and Eq. (20) is used to obtain:

$$P_{re} = ck^2\int_0^\infty d\rho\rho\int_0^{2\pi} d\phi\int_0^\infty d\eta\eta J_m(\eta\rho)\bar{a}_{x0}^{\pm}(\eta,\phi,0)\int_0^\infty d\bar{\eta}\bar{\eta}J_m(\bar{\eta}\rho)\bar{a}_{x0}^{\pm*}(\bar{\eta},\phi,0) \tag{21}$$

From the mth order Bessel transform pair, the following relation is established:

$$\int_0^\infty d\rho\rho J_m(\eta\rho)J_m(\bar{\eta}\rho) = \frac{\delta(\eta - \bar{\eta})}{\eta} \tag{22}$$

Using Eq. (22), the ρ integration in Eq. (21) is carried out; then the $\bar{\eta}$ integration is also performed. The result is

$$P_{re} = ck^2\int_0^{2\pi} d\phi\int_0^\infty d\eta\eta\bar{a}_{x0}^{\pm}(\eta,\phi,0)\overrightarrow{a}_{x0}^{\pm*}(\eta,\phi,0) \tag{23}$$

From Eq. (12), $\bar{a}_{x0}^{\pm}(\eta,\phi,0)$ is substituted into Eq. (23) and the ϕ integration is performed to yield

$$P_{re} = c\pi\varepsilon_m N_{nm}^2 \frac{1}{2^{m+2}} w_0^{4n+2m+4} \int_0^\infty d\eta\eta\eta^{4n+2m}\exp\left(-\frac{\eta^2 w_0^2}{2}\right) \tag{24}$$

where $\varepsilon_m = 2$ for $m = 0$ and $\varepsilon_m = 1$ for $m \geq 1$. The variable of integration is changed as $\tau = \eta^2 w_0^2/2$. Then, Eq. (24) simplifies as

$$P_{re} = c\pi\varepsilon_m \frac{N_{nm}^2 w_0^2}{4} 2^{2n} \int_0^\infty d\tau\tau^{2n+m}\exp(-\tau) = \frac{c\pi w_0^2 \varepsilon_m 2^{2n}}{4}(2n+m)!N_{nm}^2 \tag{25}$$

The normalization constant is chosen as

$$N_{nm} = \left[\frac{8}{c\pi w_0^2 \varepsilon_m 2^{2n}(2n+m)!}\right]^{1/2} \tag{26}$$

with the result that

$$P_{re} = 2 \tag{27}$$

The time-averaged power transported by the paraxial beam in the $\pm z$ direction is denoted by $P_{CL}^\pm$. By symmetry, $P_{CL}^+ = P_{CL}^-$. The real power P_{re} is the time-averaged power created by the current source; one half of this power is transported in the $+z$ direction and the other half in the $-z$ direction. It follows from Eq. (27) that

$$P_{CL}^+ = P_{CL}^- = \frac{1}{2}P_{re} = 1 \tag{28}$$

The normalization constant N_{nm} is chosen such that the time-averaged power $P_{CL}^\pm$ transported by the paraxial beam in the $\pm z$ direction is given by $P_{CL}^\pm = 1$ W.

2 Complex space source

In order to obtain the source in the complex space for the full-wave generalization of the complex-argument Laguerre–Gauss beam, the search for the source is made at $|z| - ib = 0$. Then, the complex space source is found from Eq. (13) as

$$C_{s,nm}(\rho,\phi,z) = \frac{N_{nm}}{ik}(-1)^n 2^{-m/2}\frac{w_0^2}{2}w_0^{2n+m}\cos m\phi\, I(\rho,n,m) \quad \text{for } |z| - ib = 0 \tag{29}$$

where

$$I(\rho,n,m) = \int_0^\infty d\eta\eta J_m(\eta\rho)\eta^{2n+m} \tag{30}$$

First, the cylindrically symmetric case corresponding to $m = 0$ is considered. The transverse Laplacian operator is given by

$$\nabla_{tm}^2 = \frac{\partial^2}{\partial\rho^2} + \frac{1}{\rho}\frac{\partial}{\partial\rho} - \frac{m^2}{\rho^2} \tag{31}$$

The differential equation satisfied by $J_m(\eta\rho)$ is obtained as

$$(\nabla_{tm}^2 + \eta^2)J_m(\eta\rho) = 0 \tag{32}$$

From Eqs. (30)–(32) and (7.38), it is found that

$$I(\rho, n, 0) = (-1)^n (\nabla_{t0}^2)^n \left[\frac{\delta(\rho)}{\rho}\right] \quad \text{for } |z| - ib = 0 \tag{33}$$

This higher-order point source in the complex space was introduced in [3] for the full-wave generalization of the cylindrically symmetric complex-argument Laguerre–Gauss beam. Second, the azimuthally varying $(m \geq 1)$ beam with the radial mode number $n = 0$ is considered. For this case, Eq. (30) leads to

$$I(\rho, 0, m) = \int_0^\infty d\eta\eta J_m(\eta\rho)\eta^m \tag{34}$$

The use of Eq. (8.18) enables us to transform Eq. (34) as

$$I(\rho, 0, m) = (-1)^m \rho^m \left(\frac{1}{\rho}\frac{\partial}{\partial\rho}\right)^m \int_0^\infty d\eta\eta J_0(\eta\rho) \tag{35}$$

Substituting Eq. (7.38) into Eq. (35) yields

$$I(\rho, 0, m) = (-1)^m \rho^m \left(\frac{1}{\rho}\frac{\partial}{\partial\rho}\right)^m \left[\frac{\delta(\rho)}{\rho}\right] \quad \text{for } |z| - ib = 0 \tag{36}$$

This higher-order point source in the complex space was introduced in [4] for the full-wave generalization of the azimuthally varying complex-argument Laguerre–Gauss beam with the radial mode number $n = 0$. The complex space source given by Eqs. (29) and (36) is the same as that introduced in Eq. (8.13), except for the normalization factor N_{0m}/ik. Bandres and Gutierrez-Vega [4] combined the operator in Eq. (36) with the source in Eq. (33) to obtain the full-wave generalization of the azimuthally varying complex-argument Laguerre–Gauss beam. To focus attention on the higher-order point source in the complex space given by Eq. (36), the source introduced by Bandres and Gutierrez-Vega [4] is discussed independently in connection with the azimuthally varying higher-order hollow full Gaussian wave in Section 8.1.

It is usual to separate out the azimuthal dependence first; then the radial eigenfunction depends on the azimuthal mode number m. It is possible to express the source entirely in terms of the transverse Laplacian operator given by Eq. (31) that also contains the azimuthal mode number m. In view of Eq. (32), Eq. (30) is recast as

$$I(\rho, n, m) = (-1)^n (\nabla_{tm}^2)^n \int_0^\infty d\eta\eta J_m(\eta\rho)\eta^m \tag{37}$$

A small length parameter ρ_{ex} is introduced. Since [6]

$$\lim_{\rho_{ex}\to 0} J_m(\eta\rho_{ex}) = \frac{\rho_{ex}^m \eta^m}{2^m m!} \tag{38}$$

Eq. (37) is transformed as

$$I(\rho, n, m) = \underset{\rho_{ex} \to 0}{\text{Lim}} \frac{2^m m!}{\rho_{ex}^m} (-1)^n (\nabla_{tm}^2)^n \int_0^\infty d\eta \eta J_m(\eta\rho) J_m(\eta\rho_{ex}) \tag{39}$$

From the mth order Bessel transform pair, another relation similar to Eq. (22) is deduced as follows:

$$\int_0^\infty d\eta \eta J_m(\eta\rho) J_m(\eta\rho_{ex}) = \frac{\delta(\rho - \rho_{ex})}{\rho} \tag{40}$$

Using Eq. (40), $I(\rho, n, m)$ is stated as a higher-order point source as

$$I(\rho, n, m) = \underset{\rho_{ex} \to 0}{\text{Lim}} \frac{2^m m!}{\rho_{ex}^m} (-1)^n (\nabla_{tm}^2)^n \frac{\delta(\rho - \rho_{ex})}{\rho} \quad \text{for } |z| - ib = 0 \tag{41}$$

The transformation of $I(\rho, n, m)$ to the form given by Eq. (41) is necessary only to ascertain that the source is highly localized in the complex space. Since only an integral expression can be obtained for the full wave, and not a closed-form solution as for the beam, the complex space source in the form stated in Eqs. (29) and (30) is sufficient for deriving the expression for the full wave.

The source given by Eqs. (29) and (30) located at $|z| - ib = 0$ generates the paraxial beam for $|z| > 0$. The same source at $|z| = 0$ produces the asymptotic limit of the paraxial beam. The paraxial equation is only an approximation for the Helmholtz equation. Therefore, the complex space source deduced from the paraxial beam, except for an excitation coefficient S_{ex}, is used for the Helmholtz equation to obtain the asymptotic limit of the full wave. Hence, the source that excites the asymptotic limit of the full-wave solution of the Helmholtz equation is obtained from Eq. (29) as

$$C_{s\infty, nm}(\rho, \phi, z) = S_{ex} \frac{N_{nm}}{ik} (-1)^n 2^{-m/2} \frac{w_0^2}{2} w_0^{2n+m} \cos m\phi \, \delta(z) I(\rho, n, m) \tag{42}$$

3 Complex-argument Laguerre–Gauss wave

Let $G_{n,m}(\rho, \phi, z)$ be the solution of the reduced Helmholtz equation [Eq. (8.2)] for the source given by Eq. (42). Then, $G_{n,m}(\rho, \phi, z)$ satisfies the following differential equation:

$$\left(\frac{\partial^2}{\partial\rho^2} + \frac{1}{\rho}\frac{\partial}{\partial\rho} - \frac{m^2}{\rho^2} + \frac{\partial^2}{\partial z^2} + k^2 \right) G_{n,m}(\rho, \phi, z) = -S_{ex} \frac{N_{nm}}{ik} (-1)^n 2^{-m/2} \frac{w_0^2}{2} w_0^{2n+m} \cos m\phi$$
$$\times \delta(z) \int_0^\infty d\eta \eta J_m(\eta\rho)\eta^{2n+m} \tag{43}$$

This differential equation is solved in the same way as Eqs. (7.44) and (8.22). The result is

$$G_{n,m}(\rho, \phi, z) = \frac{iS_{ex}}{2} \frac{N_{nm}}{ik} (-1)^n 2^{-m/2} \frac{w_0^2}{2} w_0^{2n+m} \cos m\phi \int_0^\infty d\eta \eta J_m(\eta\rho)\eta^{2n+m} \zeta^{-1} \exp(i\zeta|z|)$$
$$\tag{44}$$

where

$$\zeta = (k^2 - \eta^2)^{1/2} \tag{45}$$

$G_{n,m}(\rho, \phi, z)$ is analytically continued from $|z|$ to $|z| - ib = 0$ to obtain $A_x^{\pm}(\rho, \phi, z)$ as

$$A_x^{\pm}(\rho, \phi, z) = \frac{iS_{ex}}{2} \frac{N_{nm}}{ik} (-1)^n 2^{-m/2} \frac{w_0^2}{2} w_0^{2n+m} \cos m\phi$$

$$\times \int_0^{\infty} d\eta \eta J_m(\eta\rho) \eta^{2n+m} \zeta^{-1} \exp\left[i\zeta(|z| - ib)\right] \tag{46}$$

The paraxial approximation $(\eta^2 \ll k^2)$ of Eq. (46) is found as

$$A_{x0}^{\pm}(\rho, \phi, z) = \exp(\pm ikz) \frac{iS_{ex}}{2k} \exp(kb)$$

$$\times \frac{N_{nm}}{ik} (-1)^n 2^{-m/2} \frac{w_0^2}{2} w_0^{2n+m} \cos m\phi$$

$$\times \int_0^{\infty} d\eta \eta J_m(\eta\rho) \eta^{2n+m} \exp\left[-\frac{\eta^2 w_0^2}{4q_{\pm}^2}\right] \tag{47}$$

When Eq. (47) is compared with Eq. (13) together with Eq. (5), it follows that $A_x^{\pm}(\rho, \phi, z)$, given by Eq. (46), reduces correctly to the initially chosen paraxial beam if the excitation coefficient is chosen as

$$S_{ex} = -2ik \exp(-kb) \tag{48}$$

Substituting S_{ex} from Eq. (48) into Eq. (46) yields

$$A_x^{\pm}(\rho, \phi, z) = k \exp(-kb) \frac{N_{nm}}{ik} (-1)^n 2^{-m/2} \frac{w_0^2}{2} w_0^{2n+m} \cos m\phi$$

$$\times \int_0^{\infty} d\eta \eta J_m(\eta\rho) \eta^{2n+m} \zeta^{-1} \exp\left[i\zeta(|z| - ib)\right] \tag{49}$$

The full-wave generalization given by Eq. (49) is designated the basic full complex-argument Laguerre–Gauss wave.

4 Real and reactive powers

To evaluate the complex power, only the field components $E_x^{\pm}(\rho, \phi, z)$ and $H_y^{\pm}(\rho, \phi, z)$ are needed. From Eqs. (2.10) and (49), $H_y^{\pm}(\rho, \phi, z)$ is found as

$$H_y^{\pm}(\rho, \phi, z) = \pm \exp(-kb) N_{nm} (-1)^n 2^{-m/2} \frac{w_0^2}{2} w_0^{2n+m} \cos m\phi$$

$$\times \int_0^{\infty} d\eta \eta J_m(\eta\rho) \eta^{2n+m} \exp\left[i\zeta(|z| - ib)\right] \tag{50}$$

The electric current density induced on the $z = 0$ plane is obtained from Eq. (50) as

$$\mathbf{J}(\rho, \phi, z) = \hat{z} \times \hat{y}[H_y^+(\rho, \phi, 0) - H_y^-(\rho, \phi, 0)]\delta(z)$$

$$= -\hat{x}2\exp(-kb)N_{nm}(-1)^n 2^{-m/2}\frac{w_0^2}{2}w_0^{2n+m}$$

$$\times \cos m\phi \,\delta(z)\int_0^\infty d\bar{\eta}\bar{\eta}J_m(\bar{\eta}\rho)\bar{\eta}^{2n+m}\exp(\bar{\zeta}b) \tag{51}$$

where $\bar{\zeta}$ is the same as ζ with η replaced by $\bar{\eta}$. The complex power associated with the electric current source is given by [see Eq. (D18)]

$$P_C = P_{re} + iP_{im} = -\frac{c}{2}\int_0^\infty d\rho\rho\int_0^{2\pi} d\phi\int_{-\infty}^\infty dz\mathbf{E}^\pm(\rho, \phi, z)\cdot\mathbf{J}^*(\rho, \phi, z) \tag{52}$$

The integration with respect to z is carried out. From Eqs. (51)–(52), P_C is obtained as

$$P_C = P_{re} + iP_{im} = -\frac{c}{2}\int_0^\infty d\rho\rho\int_0^{2\pi} d\phi E_x^\pm(\rho, \phi, 0)J_x^*(\rho, \phi, 0) \tag{53}$$

where $J_x(\rho, \phi, 0)$ is the x component of $\mathbf{J}(\rho, \phi, z)$ divided by $\delta(z)$. $J_x(\rho, \phi, 0)$ depends on ϕ only as $\cos m\phi$. The integration with respect to ϕ in Eq. (53) shows that only the parts of $E_x^\pm(\rho, \phi, 0)$ that depend on ϕ as $\cos m\phi$ give rise to nonvanishing contributions to P_C. As is shown in the following, $E_x^\pm(\rho, \phi, 0)$ has three types of dependence on ϕ, namely $\cos m\phi$, $\cos(m + 2)\phi$, and $\cos(m - 2)\phi$. For $m \neq 1$, the $\cos(m + 2)\phi$ and $\cos(m - 2)\phi$ terms can be omitted since they do not contribute to the complex power P_C. The mode number $m = 1$ is a singular case, since the terms with the $\cos m\phi$ and $\cos(m - 2)\phi$ dependencies are the same. Hence, for $m = 1$, only the $\cos(m + 2)\phi$ term in $E_x^\pm(\rho, \phi, 0)$ is omitted.

Equation (2.9) shows that $E_x^\pm(\rho, \phi, z)$ is related to $A_x^\pm(\rho, \phi, z)$ as

$$E_x^\pm(\rho, \phi, z) = ik\left(1 + \frac{1}{k^2}\frac{\partial^2}{\partial x^2}\right)A_x^\pm(\rho, \phi, z) \tag{54}$$

Since $x = \rho\cos\phi$ and $y = \rho\sin\phi$, it is found that

$$\frac{\partial}{\partial x} = \cos\phi\frac{\partial}{\partial\rho} - \frac{\sin\phi}{\rho}\frac{\partial}{\partial\phi} \tag{55}$$

Therefore, from Eq. (55), it is proved that

$$\frac{\partial^2 f(\rho, \phi)}{\partial x^2} = \cos^2\phi\frac{\partial^2 f(\rho, \phi)}{\partial\rho^2} + \frac{\sin^2\phi}{\rho}\frac{\partial f(\rho, \phi)}{\partial\rho}$$

$$+ \frac{\sin^2\phi}{\rho^2}\frac{\partial^2 f(\rho, \phi)}{\partial\phi^2} - \frac{\sin 2\phi}{\rho}\frac{\partial^2 f(\rho, \phi)}{\partial\rho\partial\phi} + \frac{\sin 2\phi}{\rho^2}\frac{\partial f(\rho, \phi)}{\partial\phi} \tag{56}$$

A convenient way of evaluating P_C given by Eq. (53) is by the use of Fourier integral representations such as that stated in Eq. (49) for $A_x^\pm(\rho, \phi, z)$. The ρ and ϕ dependence of $A_x^\pm(\rho, \phi, z)$ is of the general form $J_m(\eta\rho)\cos m\phi$. Therefore, we set

$$f(\rho, \phi) = J_m(\eta\rho)\cos m\phi \tag{57}$$

Equation (57) is substituted into Eq. (56), the differentiations with respect to ϕ are carried out, the products of trigonometric terms are expressed as the sum of trigonometric terms, and the differential equation satisfied by $J_m(\eta\rho)$ [see Eqs. (31) and (32)] is used to simplify the terms containing $J_m(\eta\rho)$. Thus, Eq. (56) is transformed and expressed as

$$
\frac{\partial^2 f(\rho,\phi)}{\partial x^2} = -\frac{\eta^2}{2} J_m(\eta\rho)\cos m\phi + \frac{1}{4}\left[-\eta^2 J_m(\eta\rho) \right.
$$

$$
\left. -\frac{2(m+1)}{\rho}\frac{\partial J_m(\eta\rho)}{\partial\rho} + \frac{2m(m+1)}{\rho^2} J_m(\eta\rho) \right]\cos(m+2)\phi
$$

$$
+\frac{1}{4}\left[-\eta^2 J_m(\eta\rho) + \frac{2(m-1)}{\rho}\frac{\partial J_m(\eta\rho)}{\partial\rho} + \frac{2m(m-1)}{\rho^2} J_m(\eta\rho) \right]\cos(m-2)\phi \quad (58)
$$

As explained before, for $m \neq 1$, the $\cos(m+2)\phi$ and $\cos(m-2)\phi$ terms are omitted since they do not contribute to the complex power, yielding that

$$
\frac{\partial^2 f(\rho,\phi)}{\partial x^2} = -\frac{\eta^2}{2} J_m(\eta\rho)\cos m\phi \quad \text{for } m \neq 1 \tag{59}
$$

For $m = 1$, only the $\cos(m+2)\phi$ term is omitted. The $\cos(m-2)\phi$ simplifies as

$$
-\frac{\eta^2}{4} J_m(\eta\rho)\cos m\phi \quad \text{for } m = 1
$$

The $\cos m\phi$ and $\cos(m-2)\phi$ terms are combined, with the result that

$$
\frac{\partial^2 f(\rho,\phi)}{\partial x^2} = -\frac{3\eta^2}{4} J_m(\eta\rho)\cos m\phi \quad \text{for } m = 1 \tag{60}
$$

The results stated in Eqs. (59) and (60) are combined to obtain for all m that

$$
\frac{\partial^2 f(\rho,\phi)}{\partial x^2} = -\frac{\gamma_m \eta^2}{2} J_m(\eta\rho)\cos m\phi \tag{61}
$$

where

$$
\gamma_m = 1 \quad \text{for } m \neq 1 \quad \text{and } \gamma_m = 3/2 \quad \text{for } m = 1 \tag{62}
$$

$E_x^\pm(\rho,\phi,z)$ is determined by the use of Eqs. (49), (54), (57), and (61) as

$$
E_x^\pm(\rho,\phi,z) = k\exp(-kb)N_{nm}(-1)^n 2^{-m/2}\frac{w_0^2}{2} w_0^{2n+m}\cos m\phi
$$

$$
\times \int_0^\infty d\eta\,\eta J_m(\eta\rho)\left(1 - \frac{\gamma_m\eta^2}{2k^2}\right)\eta^{2n+m}\zeta^{-1}\exp\left[i\zeta(|z| - ib)\right] \tag{63}
$$

Substituting $E_x^{\pm}(\rho,\phi,0)$ from Eq. (63) and $J_x(\rho,\phi,0)$ from Eq. (51) into Eq. (53) yields the complex power as

$$
P_C = -\frac{c}{2}\int_0^{\infty} d\rho\rho \int_0^{2\pi} d\phi k \, \exp\left(-kb\right)N_{nm}(-1)^n 2^{-m/2}
$$

$$
\times \frac{w_0^2}{2}w_0^{2n+m}\cos m\phi \left[\int_0^{\infty} d\eta\eta J_m(\eta\rho)\right.
$$

$$
\times \left(1 - \frac{\gamma_m\eta^2}{2k^2}\right)\eta^{2n+m}\zeta^{-1}\exp\left(\zeta b\right)\bigg]
$$

$$
\times (-2)\exp\left(-kb\right)N_{nm}(-1)^n 2^{-m/2}\frac{w_0^2}{2}w_0^{2n+m}
$$

$$
\times \cos m\phi \int_0^{\infty} d\bar{\eta}\bar{\eta}J_m(\bar{\eta}\rho)\bar{\eta}^{2n+m}\exp\left(\bar{\zeta}^* b\right) \tag{64}
$$

The ϕ integration is carried out; the result is $\pi\varepsilon_m$. Equation (22) is used to perform the ρ integration to obtain $\delta(\eta - \bar{\eta})/\eta$. Then, the $\bar{\eta}$ integration is performed. The result is

$$
P_C = ck\,\exp(-2kb)N_{nm}^2 2^{-m}\frac{w_0^4}{4}w_0^{4n+2m}\pi\varepsilon_m
$$

$$
\times \int_0^{\infty} d\eta\eta\left(1 - \frac{\gamma_m\eta^2}{2k^2}\right)\eta^{4n+2m}\zeta^{-1}\exp\left[b(\zeta + \zeta^*)\right] \tag{65}
$$

The reactive power P_{im}, the imaginary part of P_C, arises from η in the range $k < \eta < \infty$, where ζ is imaginary. Hence, from Eqs. (45) and (65), P_{im} is determined as

$$
P_{im} = -ck\,\exp\left(-2kb\right)N_{nm}^2 2^{-m}\frac{w_0^4}{4}w_0^{4n+2m}\pi\varepsilon_m
$$

$$
\times \int_k^{\infty} d\eta\eta\left(1 - \frac{\gamma_m\eta^2}{2k^2}\right)\eta^{4n+2m}(\eta^2 - k^2)^{-1/2} \tag{66}
$$

Changing the variable of integration as given by $\eta^2 = k^2(1 + \tau^2)$, the integral in Eq. (66) is transformed as

$$
I_{k\infty} = k^{4n+2m+1}\int_0^{\infty} d\tau\left(1 - \frac{\gamma_m}{2} - \frac{\gamma_m}{2}\tau^2\right)(1 + \tau^2)^{2n+m} = \infty \tag{67}
$$

The value of the integral is infinite, as indicated in Eq. (67). Therefore,

$$
P_{im} = \infty \tag{68}
$$

The reactive power P_{im} associated with the basic full complex-argument Laguerre–Gauss wave is infinite. This result is to be expected since the complex space source [Eq. (41)] causing the full wave is a higher-order point source. It is necessary to search and find in the complex space a source distribution, instead of a point source, to obtain finite values for the reactive power in the same manner as was carried out for the fundamental Gaussian wave by the introduction of an analytic continuation different from $|z|$ to $|z| - ib$.

The real power P_{re}, the real part of P_C, arises from η in the range $0 < \eta < k$, where ζ is real. Therefore, from Eqs. (45) and (65), P_{re} is found as

$$P_{re} = \frac{2k \exp(-2kb) w_0^{4n+2m+2}}{2^{2n+m}(2n+m)!}$$

$$\times \int_0^k d\eta \left(1 - \frac{\gamma_m \eta^2}{2k^2}\right) \eta^{4n+2m} \zeta^{-1} \exp(2\zeta b) \tag{69}$$

The value of N_{nm} is substituted from Eq. (26) in obtaining Eq. (69). Consequently, the normalization of the wave amplitude is such that the real power in the corresponding paraxial beam is $P_{re} = 2$ W, as given by Eq. (27). The integration variable is changed as $\eta = k\sin\theta$. Then, Eq. (69) simplifies as

$$P_{re} = \frac{2(kw_0)^{4n+2m+2}}{2^{2n+m}(2n+m)!} \int_0^{\pi/2} d\theta \sin\theta$$

$$\times \left(1 - \frac{\gamma_m \sin^2\theta}{2}\right)(\sin\theta)^{4n+2m} \exp\left[-k^2 w_0^2 (1 - \cos\theta)\right] \tag{70}$$

For the special case of $n = 0$ and $m = 0$, Eq. (70) correctly reproduces the real power of the basic full Gaussian wave treated in Chapter 4.

The real power depends on kw_0, n, and m. The real power P_{re} is shown in Fig. 1 as a function of kw_0 for $0.2 < kw_0 < 10$ and for five different sets of mode numbers. P_{re} for $n = 1$, $m = 0$ is identical to that for $n = 0$, $m = 2$. The real power increases monotonically, approaching the limiting value of the paraxial beam as kw_0 is increased. For fixed

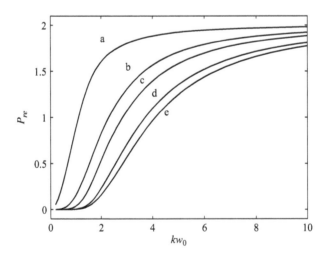

Fig. 1. Real power P_{re} as a function of kw_0 for $0.2 < kw_0 < 10$ and (a) $n = 0$, $m = 0$, (b) $n = 0$, $m = 1$, (c) $n = 0$, $m = 2$, (d) $n = 0$, $m = 3$, and (e) $n = 1$, $m = 1$. P_{re} for $n = 1$, $m = 0$ is the same as that for $n = 0$, $m = 2$. The normalization is such that the real power in the corresponding paraxial beam is 2 W.

kw_0, the real power decreases as the order of the mode increases. For example, Fig. 1 shows that for $n = 0$, the real power P_{re} decreases as m increases from 0 to 3.

References

1. T. Takenaka, M. Yokota, and O. Fukumitsu, "Propagation for light beams beyond the paraxial approximation," *J. Opt. Soc. Am. A* **2**, 826–829 (1985).

2. S. R. Seshadri, "Complex-argument Laguerre-Gauss beams: transport of mean-squared beam width," *Appl. Opt.* **44**, 7339–7343 (2005).

3. S. R. Seshadri, "Virtual source for a Laguerre-Gauss beam," *Opt. Lett.* **27**, 1872–1874 (2002).

4. M. A. Bandres and J. C. Gutierrez-Vega, "Higher-order complex source for elegant Laguerre-Gaussian waves," *Opt. Lett.* **29**, 2213–2215 (2004).

5. I. S. Gradshteyn and I. M. Ryzhik, *Tables of Integrals, Series and Products* (Academic Press, New York, 1965).

6. W. Magnus and F. Oberhettinger, *Functions of Mathematical Physics* (Chelsea Publishing Company, New York, 1954), p. 16.

CHAPTER 10

Basic full real-argument Laguerre–Gauss wave

The paraxial wave equation in the cylindrical coordinate system, in the same manner as the complex-argument Laguerre–Gauss beams, has another series of higher-order solutions known as the real-argument Laguerre–Gauss beams [1]. These higher-order Gaussian beams, which form a complete set of eigenfunctions, are characterized by the radial mode number n and the azimuthal mode number m. The fundamental Gaussian beam is the lowest-order $(n = 0, m = 0)$ mode in this set. For $n = 0$, the real-argument Laguerre–Gauss beams become identical to the complex-argument Laguerre–Gauss beams. Therefore, for the real-argument Laguerre–Gauss beams as well, the higher-order hollow Gaussian beams treated in Chapter 8 correspond to $n = 0$. In this chapter, the various aspects of the real-argument Laguerre–Gauss beams are discussed. As for the complex-argument Laguerre–Gauss beams, as well as for the real-argument Laguerre–Gauss beams, the reactive power is zero. The source in the complex space required for the full-wave generalization of the real-argument Laguerre–Gauss beams is derived. The basic full real-argument Laguerre–Gauss wave generated by the complex space source is determined. The real and the reactive powers of the basic full higher-order wave are obtained. The source in the complex space is shown to be a series of higher-order point sources that are similar to those obtained for the complex-argument Laguerre–Gauss beams. Consequently, as is to be expected, the reactive power is infinite, as for the basic full complex-argument Laguerre–Gauss wave. The general characteristics of the real power are investigated. The real power increases, approaching the limiting value of the paraxial beam as kw_0 is increased. For sufficiently large and fixed kw_0, in general, the real power decreases as the mode order increases.

1 Real-argument Laguerre–Gauss beam

1.1 Paraxial beam

The source surface electric current located on the plane $z = 0$ causes the waves that propagate in the $+z$ direction in $z > 0$ and in the $-z$ direction in $z < 0$. The electric current is in the x direction and the x component of the magnetic vector potential, $A_x^\pm(\rho, \phi, z)$, is generated, where $\pm$ indicates that the propagation is in the $\pm z$ direction. $A_x^\pm(\rho, \phi, z)$ is continuous across the $z = 0$ plane. The paraxial approximation of $A_x^\pm(\rho, \phi, z)$ is given by $A_{x0}^\pm(\rho, \phi, z)$, where the additional subscript 0 is used to denote paraxial. The slowly varying amplitude $a_{x0}^\pm(\rho, \phi, z)$ is introduced as follows:

$$A_{x0}^\pm(\rho, \phi, z) = \exp(\pm ikz) a_{x0}^\pm(\rho, \phi, z) \tag{1}$$

where $\exp(\pm ikz)$ is the rapidly varying plane-wave phase term. For the excitation of a linearly polarized real-argument Laguerre–Gauss electromagnetic beam, the required x component of the paraxial approximation of the magnetic vector potential at the secondary source or the input ($z = 0$) plane is specified as

$$A_{x0}^\pm(\rho, \phi, 0) = a_{x0}^\pm(\rho, \phi, 0) = \frac{N_{nm}}{ik} \cos m\phi \left(\frac{2\rho^2}{w_0^2}\right)^{m/2} L_n^m\left(\frac{2\rho^2}{w_0^2}\right) \exp\left(-\frac{\rho^2}{w_0^2}\right) \tag{2}$$

where N_{nm} is the normalization constant, k is the wavenumber, n is the radial mode number, and m is the azimuthal mode number. It is possible for the normalization constant to depend on n and m; therefore, the subscripts n and m are included for N. The mode numbers n and m are positive integers starting with 0. $L_n^m(\ )$ is the associated or generalized Laguerre polynomial. For $m = 0$, $L_n^m(\ )$ is denoted by $L_n(\ )$.

Since the medium in $|z| > 0$ is cylindrically symmetrical, the ϕ dependence of $A_{x0}^\pm(\rho, \phi, z)$ and $a_{x0}^\pm(\rho, \phi, z)$ remains unchanged on propagation. Therefore, $A_{x0}^\pm(\rho, \phi, z)$ satisfies the reduced Helmholtz equation:

$$\left(\frac{\partial^2}{\partial \rho^2} + \frac{1}{\rho}\frac{\partial}{\partial \rho} - \frac{m^2}{\rho^2} + \frac{\partial^2}{\partial z^2} + k^2\right) A_{x0}^\pm(\rho, \phi, z) = 0 \tag{3}$$

Substituting Eq. (1) into Eq. (3) leads to the reduced paraxial wave equation satisfied by $a_{x0}^\pm(\rho, \phi, z)$ as

$$\left(\frac{\partial^2}{\partial \rho^2} + \frac{1}{\rho}\frac{\partial}{\partial \rho} - \frac{m^2}{\rho^2} \pm 2ik\frac{\partial}{\partial z}\right) a_{x0}^\pm(\rho, \phi, z) = 0 \tag{4}$$

The mth order Bessel transform representation of $a_{x0}^\pm(\rho, \phi, z)$ is substituted into Eq. (4). Then, $\bar{a}_{x0}^\pm(\eta, \phi, z)$ satisfies the differential equation:

$$\left(\frac{\partial}{\partial z} \pm \frac{\eta^2 w_0^2}{4}\frac{i}{b}\right) \bar{a}_{x0}^\pm(\eta, \phi, z) = 0 \tag{5}$$

where $b = \frac{1}{2}kw_0^2$, and $\bar{a}_{x0}^\pm(\eta, \phi, z)$ is the mth order Bessel transform of $a_{x0}^\pm(\rho, \phi, z)$. The solution of Eq. (5) is

$$\bar{a}_{x0}^\pm(\eta, \phi, z) = \bar{a}_{x0}^\pm(\eta, \phi, 0) \exp\left(-\frac{\eta^2 w_0^2}{4}\frac{i|z|}{b}\right) \tag{6}$$

The mth order Bessel transform of $A_{x0}^{\pm}(\rho,\phi,0)$ and $a_{x0}^{\pm}(\rho,\phi,0)$ is expressed as

$$\overline{A}_{x0}^{\pm}(\eta,\phi,0) = \overline{a}_{x0}^{\pm}(\eta,\phi,0) = \frac{N_{nm}}{ik}\cos m\phi \frac{2^{m/2}}{w_0^m}\int_0^\infty d\rho\rho J_m(\eta\rho)\rho^m L_n^m\left(\frac{2\rho^2}{w_0^2}\right)\exp\left(-\frac{\rho^2}{w_0^2}\right) \quad (7)$$

The integral in Eq. (7) is evaluated by the use of the formula 7.421.4 in [2] as given by

$$\int_0^\infty dx x J_m(yx)x^m L_n^m(\alpha x^2)\exp\left(-\beta x^2\right) = \frac{1}{2^{m+1}}\frac{(\beta-\alpha)^n}{\beta^{n+m+1}}y^m L_n^m\left[\frac{\alpha y^2}{4\beta(\alpha-\beta)}\right]\exp\left(-\frac{y^2}{4\beta}\right) \quad (8)$$

We set

$$\alpha = 2/w_0^2 \quad \text{and} \quad \beta = 1/w_0^2 \quad (9)$$

From Eq. (9), it is found that

$$(\beta-\alpha)^n = (-1)^n/w_0^{2n}, \quad \beta^{n+m+1} = w_0^{-2(n+m+1)} \quad (10)$$

$$\frac{1}{4\beta} = \frac{w_0^2}{4}, \quad \text{and} \quad \frac{\alpha}{4\beta(\alpha-\beta)} = \frac{w_0^2}{2} \quad (11)$$

The integration in Eq. (7) is performed by the use of Eqs. (8)–(11), and the result is

$$\overline{a}_{x0}^{\pm}(\eta,\phi,0) = \frac{N_{nm}}{ik}\frac{\cos m\phi}{2^{m/2+1}}(-1)^n w_0^{m+2}\eta^m L_n^m\left(\frac{w_0^2\eta^2}{2}\right)\exp\left(-\frac{w_0^2\eta^2}{4}\right) \quad (12)$$

The result of substituting Eq. (12) into Eq. (6) is

$$\overline{a}_{x0}^{\pm}(\eta,\phi,z) = \frac{N_{nm}}{ik}\frac{\cos m\phi}{2^{m/2+1}}(-1)^n w_0^{m+2}\eta^m L_n^m\left(\frac{w_0^2\eta^2}{2}\right)\exp\left(-\frac{w_0^2\eta^2}{4q_\pm^2}\right) \quad (13)$$

where

$$q_\pm = (1 + i|z|/b)^{-1/2} \quad (14)$$

The inverse Bessel transform of Eq. (13) is substituted into Eq. (1), with the result that

$$A_{x0}^{\pm}(\rho,\phi,z) = \frac{N_{nm}}{ik}\exp\left(\pm ikz\right)\frac{\cos m\phi}{2^{m/2+1}}(-1)^n w_0^{m+2}$$
$$\times \int_0^\infty d\eta\eta J_m(\eta\rho)\eta^m L_n^m\left(\frac{w_0^2\eta^2}{2}\right)\exp\left(-\frac{w_0^2\eta^2}{4q_\pm^2}\right) \quad (15)$$

The integral in Eq. (15) is also evaluated by the use of formula 7.421.4 in [2], as reproduced in Eq. (8). Now we set

$$\alpha = w_0^2/2 \quad \text{and} \quad \beta = w_0^2/4q_\pm^2 \quad (16)$$

From Eq. (16), it is obtained that

$$(\beta - \alpha)^n = (-1)^n \left(\frac{w_0}{2q_\mp}\right)^{2n}, \quad \beta^{n+m+1} = \left(\frac{w_0}{2q_\pm}\right)^{2n+2m+2} \tag{17}$$

$$\frac{1}{4\beta} = \frac{q_\pm^2}{w_0^2}, \quad \text{and} \quad \frac{\alpha}{4\beta(\alpha - \beta)} = \frac{2q_\mp^2 q_\pm^2}{w_0^2} \tag{18}$$

The integration in Eq. (15) is performed using Eqs. (8) and (16)–(18) to yield that

$$A_{x0}^\pm(\rho, \phi, z) = \frac{N_{nm}}{ik} \exp(\pm ikz) \cos m\phi \, \frac{q_\pm^{2n+m+2}}{q_\mp^{2n+m}}$$

$$\times \left(\frac{2q_\mp^2 q_\pm^2 \rho^2}{w_0^2}\right)^{m/2} L_n^m\left(\frac{2q_\mp^2 q_\pm^2 \rho^2}{w_0^2}\right) \exp\left(-\frac{q_\pm^2 \rho^2}{w_0^2}\right) \tag{19}$$

Equation (19) is the vector potential governing the real-argument Laguerre–Gauss beam.

The real-argument Laguerre–Gauss beam has a closed-form Bessel transform representation, as given by Eq. (13). For the complex-argument Laguerre–Gauss beam, the associated Laguerre polynomial appears in the spatial domain description [Eq. (9.16)] but is absent in the Bessel transform or wavenumber representation [Eq. (9.13)]. Therefore, the wavenumber representation is expected to be simpler than the spatial domain description. But for the real-argument Laguerre–Gauss beam, the associated Laguerre polynomial appears both in the spatial domain [Eq. (19)] and in the wavenumber representations [Eq. (13)]. Consequently, for the real-argument Laguerre–Gauss beam, the wavenumber representation is not expected to provide any significant advantages over the spatial domain description.

1.2 Complex power

$E_{x0}^\pm(\rho, \phi, z)$ and $H_{y0}^\pm(\rho, \phi, z)$ are the only electromagnetic field components associated with the paraxial beam [see Eq. (1.12)]. $A_{x0}^\pm(\rho, \phi, z)$ and $E_{x0}^\pm(\rho, \phi, z) [= ikA_{x0}^\pm(\rho, \phi, z)]$ are continuous, but $H_{y0}^\pm(\rho, \phi, z) [= \pm ikA_{x0}^\pm(\rho, \phi, z)]$ is discontinuous across the plane $z = 0$. The electric current density induced on the $z = 0$ plane as a consequence of the discontinuity in $H_{y0}^\pm(\rho, \phi, z)$ is given by

$$\mathbf{J}_0(\rho, \phi, z) = \hat{z} \times \hat{y}[H_y^+(\rho, \phi, 0) - H_y^-(\rho, \phi, 0)]\delta(z)$$

$$= -\hat{x}2ikA_{x0}^\pm(\rho, \phi, 0)\delta(z) \tag{20}$$

The complex power is obtained by the use of Eqs. (1) and (9.19) as

$$P_C = P_{re} + iP_{im} = ck^2 \int_0^\infty d\rho\rho \int_0^{2\pi} d\phi a_{x0}^\pm(\rho, \phi, 0) a_{x0}^{\pm*}(\rho, \phi, 0) \tag{21}$$

P_C given by Eq. (21) is real. Therefore,

$$P_{im} = 0 \tag{22}$$

The reactive power of the real-argument Laguerre–Gauss beam vanishes.

By the use of Eq. (19) in Eq. (21), P_{re} is expressed as

$$P_{re} = cN_{nm}^2 \int_0^{2\pi} d\phi \cos^2 m\phi \int_0^\infty d\rho\rho$$

$$\times \left(\frac{2\rho^2}{w_0^2}\right)^m \left[L_n^m\left(\frac{2\rho^2}{w_0^2}\right)\right]^2 \exp\left(-\frac{2\rho^2}{w_0^2}\right) \tag{23}$$

Carrying out the ϕ integration and changing the variable of the ρ integration as $u = 2\rho^2/w_0^2$ leads to

$$P_{re} = cN_{nm}^2 \pi\varepsilon_m \frac{w_0^2}{4} \int_0^\infty du u^m [L_n^m(u)]^2 \exp(-u) \tag{24}$$

where $\varepsilon_m = 2$ for $m = 0$ and $\varepsilon_m = 1$ for $m \geq 1$. The radial eigenfunction of the real-argument Laguerre–Gauss beam or the associated Laguerre polynomial satisfies the following orthogonality relation {formula 7.414.3 in [2]}:

$$\int_0^\infty dx x^m L_n^m(x) L_p^m(x) \exp(-x) = \delta_{np} \frac{(n+m)!}{n!} \tag{25}$$

where $\delta_{np} = 0$ for $n \neq p$ and $\delta_{np} = 1$ for $n = p$. Using Eq. (25) in Eq. (24) permits P_{re} to be simplified as

$$P_{re} = cN_{nm}^2 \pi\varepsilon_m \frac{w_0^2}{4} \frac{(n+m)!}{n!} \tag{26}$$

The normalization constant is set as

$$N_{nm} = \left[\frac{8n!}{c\pi\varepsilon_m w_0^2 (n+m)!}\right]^{1/2} \tag{27}$$

Therefore,

$$P_{re} = 2 \tag{28}$$

The time-averaged power transported by the paraxial beam in the $\pm z$ direction is denoted by $P_{RL}^\pm$. By symmetry, $P_{RL}^+ = P_{RL}^-$. The real power P_{re} is the time-averaged power created by the current source; one half of this power is transported in the $+z$ direction and the other half in the $-z$ direction. It follows from Eq. (28) that

$$P_{RL}^+ = P_{RL}^- = \frac{1}{2} P_{re} = 1 \tag{29}$$

The normalization constant N_{nm} is chosen such that the time-averaged power $P_{RL}^\pm$ transported by the paraxial beam in the $\pm z$ direction is given by $P_{RL}^\pm = 1$ W.

For $n = 0$, $L_0^m(\ ) = 1$, and Eq. (19) becomes identical to Eq. (9.16). Also, N_{0m} given by Eq. (27) becomes the same as that given by Eq. (9.26). Therefore, for $n = 0$, the real-argument Laguerre–Gauss beam becomes identical to the corresponding complex-argument Laguerre–Gauss beam.

2 Real-argument Laguerre–Gauss wave

$L_n^m(x)$ has the following series expansion {formula 8.970.1 in [2]}:

$$L_n^m(x) = \sum_{r=0}^{r=n} \frac{(-1)^r (n+m)! x^r}{(n-r)! r! (m+r)!} \tag{30}$$

From Eq. (15), by the use of Eqs. (1) and (30), it is found that

$$a_{x0}^\pm(\rho,\phi,z) = \frac{N_{nm}}{ik} \frac{\cos m\phi}{2^{m/2+1}} (-1)^n w_0^{m+2} \sum_{r=0}^{r=n} \frac{(n+m)!(-1)^r w_0^{2r}}{2^r (n-r)! r! (m+r)!}$$

$$\times \int_0^\infty d\eta \eta J_m(\eta\rho) \eta^{2r+m} \exp\left(-\frac{w_0^2 \eta^2}{4q_\pm^2}\right) \tag{31}$$

The search for the complex space source is made at $|z| - ib = 0$, where $q_\pm^{-2} = 0$. Then, the complex space source is determined from Eq. (31) as

$$C_{s,nm}(\rho,\phi,z) = \frac{N_{nm}}{ik} \frac{\cos m\phi}{2^{m/2+1}} (-1)^n w_0^{m+2}$$

$$\times \sum_{r=0}^{r=n} \frac{(n+m)!(-1)^r w_0^{2r}}{2^r (n-r)! r! (m+r)!} I(\rho,r,m) \tag{32}$$

where

$$I(\rho,r,m) = \int_0^\infty d\eta \eta J_m(\eta\rho) \eta^{2r+m} \tag{33}$$

It follows from Eqs. (32), (33), (9.29), and (9.30) that, for the real-argument Laguerre–Gauss beam, the complex space source consists of a series of $(n+1)$ terms, and each term of the series is similar to the complex space source associated with the complex-argument Laguerre–Gauss beam. By the use of Eqs. (9.30) and (9.41), Eq. (33) is expressed as

$$I(\rho,r,m) = \lim_{\rho_{ex}\to 0} \frac{2^m m!}{\rho_{ex}^m} (-1)^r (\nabla_{tm}^2)^r \frac{\delta(\rho-\rho_{ex})}{\rho} \quad \text{for } |z| - ib = 0 \tag{34}$$

where

$$\nabla_{tm}^2 = \frac{\partial^2}{\partial\rho^2} + \frac{1}{\rho}\frac{\partial}{\partial\rho} - \frac{m^2}{\rho^2} \tag{35}$$

The transformation of $I(\rho, r, m)$ to the form given by Eq. (34) reveals that the complex space source given by Eq. (32) is a series of $(n+1)$ higher-order point sources, all located at $|z| - ib = 0$ and $\rho \to 0$. A closed-form solution [Eq. (19)] is obtained for the paraxial beam. But for the full wave, only an integral expression can be determined. Therefore, the complex space source in the form stated in Eqs. (32) and (33) is sufficient to derive the expression for the full wave. The series form of $L_n^m(w_0^2\eta^2/2)$ is replaced by $L_n^m(w_0^2\eta^2/2)$ for convenience.

As explained in connection with the complex-argument Laguerre–Gauss beam, the complex space source deduced from the paraxial beam, except for an excitation coefficient S_{ex}, is used for the Helmholtz equation to determine the asymptotic limit of the full wave. Hence, the source that generates the asymptotic limit of the full-wave solution of the Helmholtz equation is obtained from Eq. (32) as

$$C_{s\infty, nm}(\rho, \phi, z) = S_{ex} \frac{N_{nm}}{ik} \frac{\cos m\phi}{2^{m/2+1}} (-1)^n w_0^{m+2}$$

$$\times \int_0^\infty d\eta \eta J_m(\eta\rho)\eta^m L_n^m\left(\frac{w_0^2\eta^2}{2}\right) \tag{36}$$

Let $G_{n,m}(\rho, \phi, z)$ be the solution of the reduced Helmholtz equation [Eq. (3)] for the source given by Eq. (36). Then, $G_{n,m}(\rho, \phi, z)$ satisfies the following differential equation:

$$\left(\frac{\partial^2}{\partial\rho^2} + \frac{1}{\rho}\frac{\partial}{\partial\rho} - \frac{m^2}{\rho^2} + \frac{\partial^2}{\partial z^2} + k^2\right) G_{n,m}(\rho, \phi, z) = -S_{ex} \frac{N_{nm}}{ik} \frac{\cos m\phi}{2^{m/2+1}} (-1)^n w_0^{m+2}$$

$$\times \int_0^\infty d\eta \eta J_m(\eta\rho)\eta^m L_n^m\left(\frac{w_0^2\eta^2}{2}\right) \tag{37}$$

This differential equation is solved in the same manner as Eqs. (7.44) and (8.22). $G_{n,m}(\rho, \phi, z)$ is expressed in terms of its mth order Bessel transform, $\overline{G}_{n,m}(\eta, \phi, z)$. Then, from Eq. (37), the differential equation satisfied by $\overline{G}_{n,m}(\eta, \phi, z)$ is obtained as

$$\left(\frac{\partial^2}{\partial z^2} + \zeta^2\right) \overline{G}_{n,m}(\eta, \phi, z) = -S_{ex} \frac{N_{nm}}{ik} \frac{\cos m\phi}{2^{m/2+1}} (-1)^n w_0^{m+2} \eta^m L_n^m\left(\frac{w_0^2\eta^2}{2}\right) \tag{38}$$

where

$$\zeta = (k^2 - \eta^2)^{1/2} \tag{39}$$

The solution of Eq. (38) is determined as

$$\overline{G}_{n,m}(\eta, \phi, z) = \frac{iS_{ex}}{2} \frac{N_{nm}}{ik} \frac{\cos m\phi}{2^{m/2+1}} (-1)^n w_0^{m+2} \eta^m L_n^m\left(\frac{w_0^2\eta^2}{2}\right) \zeta^{-1} \exp(i\zeta|z|) \tag{40}$$

The inverse mth order Bessel transform of Eq. (40) yields

$$G_{n,m}(\rho, \phi, z) = \frac{iS_{ex}}{2} \frac{N_{nm}}{ik} \frac{\cos m\phi}{2^{m/2+1}} (-1)^n w_0^{m+2} \int_0^\infty d\eta \eta J_m(\eta\rho)\eta^m$$

$$\times L_n^m\left(\frac{w_0^2\eta^2}{2}\right) \zeta^{-1} \exp(i\zeta|z|) \tag{41}$$

By analytically continuing Eq. (41) from $|z|$ to $|z| - ib$, $A_x^\pm(\rho, \phi, z)$ is determined as

$$A_x^\pm(\rho, \phi, z) = \frac{iS_{ex}}{2} \frac{N_{nm}}{ik} \frac{\cos m\phi}{2^{m/2+1}} (-1)^n w_0^{m+2} \int_0^\infty d\eta \eta J_m(\eta\rho)\eta^m$$

$$\times L_n^m \left(\frac{w_0^2\eta^2}{2}\right) \zeta^{-1} \exp\left[i\zeta(|z| - ib)\right] \tag{42}$$

The paraxial approximation ($\eta^2 \ll k^2$) of Eq. (42) is derived as

$$A_{x0}^\pm(\rho, \phi, z) = \exp(\pm ikz) \frac{iS_{ex}}{2k} \exp(kb) \frac{N_{nm}}{ik} \frac{\cos m\phi}{2^{m/2+1}} (-1)^n w_0^{m+2} \int_0^\infty d\eta \eta J_m(\eta\rho)$$

$$\times \eta^m L_n^m \left(\frac{w_0^2\eta^2}{2}\right) \exp\left(-\frac{\eta^2 w_0^2}{4q_\pm^2}\right) \tag{43}$$

A comparison of Eq. (43) with Eq. (15) shows that $A_x^\pm(\rho, \phi, z)$, given by Eq. (42), reproduces the initially chosen paraxial beam if the excitation coefficient is chosen as

$$S_{ex} = -2ik \exp(-kb) \tag{44}$$

Substituting Eq. (44) into Eq. (42) yields

$$A_x^\pm(\rho, \phi, z) = k \exp(-kb) \frac{N_{nm}}{ik} \frac{\cos m\phi}{2^{m/2+1}} (-1)^n w_0^{m+2} \int_0^\infty d\eta \eta J_m(\eta\rho)\eta^m$$

$$\times L_n^m \left(\frac{w_0^2\eta^2}{2}\right) \zeta^{-1} \exp[i\zeta(|z| - ib)] \tag{45}$$

The full-wave generalization represented by Eq. (45) is designated the basic full real-argument Laguerre–Gauss wave.

3 Real and reactive powers

The field components $E_x^\pm(\rho, \phi, z)$ and $H_y^\pm(\rho, \phi, z)$ are needed for the determination of the complex power. From Eqs. (2.10) and (45), $H_y^\pm(\rho, \phi, z)$ is obtained as

$$H_y^\pm(\rho, \phi, z) = \pm \exp(-kb) N_{nm} \frac{\cos m\phi}{2^{m/2+1}} (-1)^n w_0^{m+2} \int_0^\infty d\eta \eta J_m(\eta\rho)\eta^m$$

$$\times L_n^m \left(\frac{w_0^2\eta^2}{2}\right) \exp[i\zeta(|z| - ib)] \tag{46}$$

The electric current density induced on the $z = 0$ plane is deduced from Eq. (46) as

$$\mathbf{J}(\rho, \phi, z) = \hat{z} \times \hat{y}[H_y^+(\rho, \phi, 0) - H_y^-(\rho, \phi, 0)]\delta(z)$$

$$= -\hat{x}2 \exp(-kb) N_{nm} \frac{\cos m\phi}{2^{m/2+1}} (-1)^n w_0^{m+2}$$

$$\times \delta(z) \int_0^\infty d\bar{\eta}\bar{\eta} J_m(\bar{\eta}\rho)\bar{\eta}^m L_n^m \left(\frac{w_0^2\bar{\eta}^2}{2}\right) \exp(\bar{\zeta}b) \tag{47}$$

where $\bar{\zeta}$ is the same as ζ, with η replaced by $\bar{\eta}$. The complex power P_C associated with the electric current source is given by Eq. (D18). After the integration with respect to z is carried out, from Eq. (47), P_C is found as

$$P_C = P_{re} + iP_{im} = -\frac{c}{2} \int_0^\infty d\rho\rho \int_0^{2\pi} d\phi E_x^\pm(\rho,\phi,0) J_x^*(\rho,\phi,0) \tag{48}$$

where $J_x(\rho,\phi,0)$ is the x component of $\mathbf{J}(\rho,\phi,z)$ divided by $\delta(z)$. $J_x(\rho,\phi,0)$ depends on ϕ only as $\cos m\phi$. The integration with respect to ϕ in Eq. (48) shows that only the parts of $E_x^\pm(\rho,\phi,0)$ that depend on ϕ as $\cos m\phi$ give rise to nonvanishing contributions to P_C. If $\cos m\phi$ alone is retained, it can be shown [see Eqs. (9.55)–(9.61)] that

$$\frac{\partial^2}{\partial x^2} J_m(\eta\rho)\cos m\phi = -\frac{\gamma_m\eta^2}{2} J_m(\eta\rho)\cos m\phi \tag{49}$$

where γ_m is given by Eq. (9.62). $E_x^\pm(\rho,\phi,z)$ is derived by the use of Eqs. (45), (2.9), and (49) as

$$E_x^\pm(\rho,\phi,z) = k\exp(-kb)N_{nm}\frac{\cos m\phi}{2^{m/2+1}}(-1)^n w_0^{m+2} \int_0^\infty d\eta\eta J_m(\eta\rho)\left(1 - \frac{\gamma_m\eta^2}{2k^2}\right)\eta^m$$

$$\times L_n^m\left(\frac{w_0^2\eta^2}{2}\right)\zeta^{-1}\exp[i\zeta(|z| - ib)] \tag{50}$$

Substituting $E_x^\pm(\rho,\phi,z)$ from Eq. (50) and $J_x(\rho,\phi,z)$ from Eq. (47) into Eq. (48) leads to the complex power as

$$P_C = -\frac{c}{2}\int_0^\infty d\rho\rho \int_0^{2\pi} d\phi k\exp(-kb)N_{nm}\frac{\cos m\phi}{2^{m/2+1}}(-1)^n w_0^{m+2}$$

$$\times \int_0^\infty d\eta\eta J_m(\eta\rho)\left(1 - \frac{\gamma_m\eta^2}{2k^2}\right)\eta^m$$

$$\times L_n^m\left(\frac{w_0^2\eta^2}{2}\right)\zeta^{-1}\exp(\zeta b)$$

$$\times (-2)\exp(-kb)N_{nm}\frac{\cos m\phi}{2^{m/2+1}}(-1)^n w_0^{m+2}$$

$$\times \int_0^\infty d\bar{\eta}\bar{\eta} J_m(\bar{\eta}\rho)\bar{\eta}^m L_n^m\left(\frac{w_0^2\bar{\eta}^2}{2}\right)\exp(\bar{\zeta}^* b) \tag{51}$$

The ϕ integration is carried out, and the result is $\pi\varepsilon_m$. The ρ integration is performed to obtain $\delta(\eta - \bar{\eta})/\eta$. The $\bar{\eta}$ integration is carried out with the result that

$$P_C = c\pi\varepsilon_m k\exp(-2kb)N_{nm}^2\frac{w_0^{2m+4}}{2^{m+2}}\int_0^\infty d\eta\eta\left(1 - \frac{\gamma_m\eta^2}{2k^2}\right)\eta^{2m}$$

$$\times \left[L_n^m\left(\frac{w_0^2\eta^2}{2}\right)\right]^2 \zeta^{-1}\exp[b(\zeta + \zeta^*)] \tag{52}$$

The reactive power P_{im}, the imaginary part of P_C, arises for $k < \eta < \infty$, where ζ is imaginary. Hence, from Eqs. (39) and (52), P_{im} is determined as

$$P_{im} = -c\pi\varepsilon_m k \exp\left(-2kb\right) N_{nm}^2 \frac{w_0^{2m+4}}{2^{m+2}} \int_k^\infty d\eta\eta \left(1 - \frac{\gamma_m \eta^2}{2k^2}\right) \eta^{2m}$$

$$\times \left[L_n^m\left(\frac{w_0^2 \eta^2}{2}\right)\right]^2 (\eta^2 - k^2)^{-1/2} \tag{53}$$

The integration variable is changed as $\eta^2 = k^2(1 + \tau^2)$. Then, Eq. (53) is changed to

$$P_{im} = -c\pi\varepsilon_m \exp\left(-2kb\right) N_{nm}^2 \frac{k^{2m+2} w_0^{2m+4}}{2^{m+2}} \int_0^\infty d\tau \left(1 - \frac{\gamma_m}{2} - \frac{\gamma_m}{2}\tau^2\right)(1 + \tau^2)^m$$

$$\times \left\{L_n^m\left[\frac{k^2 w_0^2}{2}(1 + \tau^2)\right]\right\}^2 \tag{54}$$

When the polynomial expansion of $L_n^m(\ )$ is introduced, it follows that

$$P_{im} = \infty \tag{55}$$

The reactive power P_{im} associated with the basic full real-argument Laguerre–Gauss wave is infinite. The complex space source is a series of higher-order point sources, all located at $|z| - ib = 0$ and $\rho \to 0$ [see Eq. (34)]. Therefore, $P_{im} = \infty$ is a result that is to be expected. Finite values of the reactive power are obtained only when the composite source in the complex space is a series of distributive sources rather than point sources.

The real power P_{re}, the real part of P_C, arises from η in the range $0 < \eta < k$, where ζ is real. Therefore, from Eqs. (39) and (52), P_{re} is determined as

$$P_{re} = 2k \exp(-2kb) \frac{n! w_0^{2m+2}}{(m+n)! 2^m} \int_0^k d\eta\eta \left(1 - \frac{\gamma_m \eta^2}{2k^2}\right) \eta^{2m} \left[L_n^m\left(\frac{w_0^2 \eta^2}{2}\right)\right]^2 \zeta^{-1} \exp(2\zeta b) \tag{56}$$

The value of N_{nm} from Eq. (27) is substituted in obtaining Eq. (56). Consequently, the normalization of the wave amplitude is such that the real power in the corresponding paraxial beam is $P_{re} = 2$ W, as given by Eq. (28). The integration variable is changed as $\eta = k\sin\theta$. Then, Eq. (56) is transformed as

$$P_{re} = \frac{n!}{(m+n)!} \frac{(kw_0)^{2m+2}}{2^{m-1}} \int_0^{\pi/2} d\theta \sin\theta \left(1 - \frac{\gamma_m}{2}\sin^2\theta\right)$$

$$\times (\sin\theta)^{2m} \left[L_n^m\left(\frac{k^2 w_0^2}{2}\sin^2\theta\right)\right]^2 \exp[-k^2 w_0^2(1 - \cos\theta)] \tag{57}$$

For the special case of $n = 0$ and $m = 0$, Eq. (57) correctly reproduces the real power of the basic full Gaussian wave (see Chapter 4).

As pointed out before, for $n = 0$, the real-argument and the complex-argument Laguerre–Gauss beams are identical. Also, since the normalization of the amplitude is the same for both types of beams, the real power of the basic full real-argument Laguerre–Gauss wave, as given by Eq. (57), is identical to the real power of the basic full complex-argument Laguerre–Gauss wave, as given by Eq. (9.70). The real power depends on kw_0, n, and m.

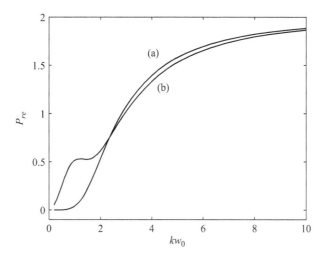

Fig. 1. Real power P_{re} as a function of kw_0 for $0.2 < kw_0 < 10$, and (a) $n = 0, m = 2$, and (b) $n = 1, m = 0$. The normalization is such that the real power in the corresponding paraxial beam is 2 W.

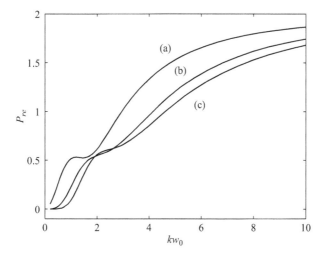

Fig. 2. Same as in Fig. 1, but for (a) $n = 1, m = 0$, (b) $n = 1, m = 1$, and (c) $n = 1, m = 2$.

The real power is shown in Figs. 1 to 4 as a function of kw_0 for $0.2 < kw_0 < 10$ and for different set of mode numbers n and m. In Fig. 1, P_{re} is shown for (a) $n = 0$, $m = 2$ and (b) $n = 1$, $m = 0$. For these two sets of n and m, P_{re} is the same for the complex-argument Laguerre–Gauss wave, but, as shown in Fig. 1, P_{re} is different for the real-argument Laguerre–Gauss wave. In Fig. 2, P_{re} is shown for $n = 1$ and $m = 0$, 1, and 2. Similarly, in Fig. 3, P_{re} is shown for $n = 2$ and $m = 0$, 1, and 2. In Fig. 4, P_{re} is shown for the cylindrically symmetric ($m = 0$) real-argument Laguerre–Gauss wave for $n = 0$, 1, and 2. In general, the real power increases, approaching the limiting value of the paraxial beam as kw_0 is increased. For

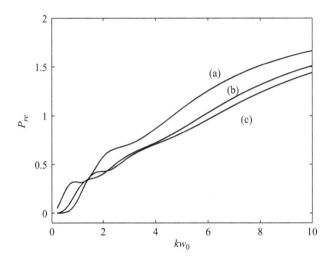

Fig. 3.   Same as in Fig. 1, but for (a) $n = 2$, $m = 0$, (b) $n = 2$, $m = 1$, and (c) $n = 2$, $m = 2$.

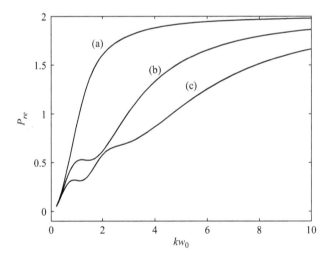

Fig. 4.   Same as in Fig. 1, but for (a) $n = 0$, $m = 0$, (b) $n = 1$, $m = 0$, and (c) $n = 2$, $m = 0$.

sufficiently large and fixed kw_0, in general, the real power decreases as the order of the mode increases. For example, Fig. 4 shows that for $m = 0$, the real power decreases as n increases from 0 to 2. As revealed by Figs. 1 and 4, for $n = 0$, the increase of P_{re} with the increase in kw_0 is monotonic. For $n \geq 1$, the variation of P_{re} with the increase in kw_0 has oscillatory behavior for small kw_0.

References

1. H. Kogelnik and T. Li, "Laser beams and resonators," *Appl. Opt.* **5**, 1550–1567 (1966).

2. I. S. Gradshteyn and I. M. Ryzhik, *Table of Integrals, Series, and Products* (Academic Press, New York, 1965).

CHAPTER 11

Basic full complex-argument Hermite–Gauss wave

The paraxial wave equation in the Cartesian coordinate system has a series of higher-order solutions known as the complex-argument Hermite–Gauss beams [1,2]. This series of eigenfunctions is a complete set. These higher-order Gaussian beams are described by the mode numbers m and n in the x and y directions, respectively. The fundamental Gaussian beam is the lowest-order ($m = 0$, $n = 0$) mode in this set. In this chapter, a treatment of the complex-argument Hermite–Gauss beams is presented. As for the other previously introduced paraxial beams, the reactive power of the complex-argument Hermite–Gauss beams vanishes. The higher-order point source in the complex space required for the full-wave generalization of the complex-argument Hermite–Gauss beams is derived. The basic full complex-argument Hermite–Gauss wave generated by the complex space source is determined [3]. The real and reactive powers of the basic full complex-argument Hermite–Gauss waves are evaluated. As for the other basic full Gaussian waves, the reactive power is infinite. The general characteristics of the real power are investigated. The real power increases, approaching the limiting value of the paraxial beam as kw_0 is increased. For a fixed kw_0, the real power decreases as the mode order increases.

1 Complex-argument Hermite–Gauss beam

1.1 Paraxial beam

The secondary source is an infinitesimally thin sheet of electric current located on the plane $z = 0$. The waves generated by the source propagate outward in the $+z$ direction in $z > 0$ and in the $-z$ direction in $z < 0$. The source electric current density is in the x direction. The generated magnetic vector potential is in the x direction and is continuous across the plane $z = 0$. For the excitation of a linearly polarized electromagnetic complex-argument Hermite–Gauss beam, the required x component of the magnetic vector potential at the secondary source plane is specified as

$$a_{x0}^{\pm}(x,y,0) = A_{x0}^{\pm}(x,y,0) = \frac{N_{mn}}{ik}(-1)^{m+n}H_m\left(\frac{x}{w_0}\right)H_n\left(\frac{y}{w_0}\right)\exp\left(-\frac{x^2+y^2}{w_0^2}\right) \qquad (1)$$

where N_{mn} is the normalization constant, k is the wavenumber, and m and n are the mode numbers in the x and y directions, respectively. The subscripts m and n on N indicate that it is possible for the normalization constant to depend on the mode orders m and n. The Hermite polynomial of order p is defined as {formula 8.950.1 in [4] and formula 22.3.10 in [5]}

$$H_p(v) = (-1)^p \exp(v^2) \frac{d^p}{dv^p} \exp(-v^2) \tag{2}$$

$$= p! \sum_{\ell=0}^{\ell=\ell_p} \frac{(-1)^\ell 2^{p-2\ell} v^{p-2\ell}}{\ell!(p-2\ell)!} \tag{3}$$

where ℓ_p is the largest integer $\leq p/2$. It follows from Eq. (2) that

$$H_0(v) = 1 \tag{4}$$

The mode numbers m and n are positive integers starting with zero.

The paraxial approximation of the x component of the magnetic vector potential is given by $A_{x0}^\pm(x,y,z)$. The additional subscript 0 is used to indicate paraxial. The rapidly varying phase is separated out as in

$$A_{x0}^\pm(x,y,z) = \exp(\pm ikz) a_{x0}^\pm(x,y,z) \tag{5}$$

The slowly varying amplitude $a_{x0}^\pm(x,y,z)$ satisfies the paraxial wave equation:

$$\left(\frac{\partial^2}{\partial x^2} + \frac{\partial^2}{\partial y^2} \pm 2ik \frac{\partial}{\partial z} \right) a_{x0}^\pm(x,y,z) = 0 \tag{6}$$

As before, the sign $\pm$ and the superscript $\pm$ are used to indicate that the propagation is in the $\pm z$ direction. The two-dimensional Fourier transform representation of $a_{x0}^\pm(x,y,z)$ is substituted into Eq. (6) to obtain

$$\left[\frac{\partial}{\partial z} \pm \frac{i}{b} \pi^2 w_0^2 \left(p_x^2 + p_y^2 \right) \right] \overline{a}_{x0}^\pm(p_x, p_y, z) = 0 \tag{7}$$

where $b = \frac{1}{2} k w_0^2$, and $\overline{a}_{x0}^\pm(p_x, p_y, z)$ is the two-dimensional Fourier transform of $a_{x0}^\pm(x,y,z)$. The solution of Eq. (7) is

$$\overline{a}_{x0}^\pm(p_x, p_y, z) = \overline{a}_{x0}^\pm(p_x, p_y, 0) \exp\left[-\pi^2 w_0^2 \left(p_x^2 + p_y^2 \right) \frac{i|z|}{b} \right] \tag{8}$$

By the use of Eq. (2), Eq. (1) is expressed as

$$a_{x0}^\pm(x,y,0) = \frac{N_{mn}}{ik} w_0^{m+n} \frac{\partial^m}{\partial x^m} \frac{\partial^n}{\partial y^n} \exp\left(-\frac{x^2 + y^2}{w_0^2} \right) \tag{9}$$

From Eqs. (1.1) and (1.3), it is found that

$$\exp\left(-\frac{x^2+y^2}{w_0^2}\right) = \pi w_0^2 \int_{-\infty}^{\infty}\int_{-\infty}^{\infty} dp_x dp_y \exp\left[-i2\pi(p_x x + p_y y)\right]\exp\left[-\pi^2 w_0^2\left(p_x^2 + p_y^2\right)\right]$$

(10)

Substituting Eq. (10) into Eq. (9) yields

$$a_{x0}^{\pm}(x,y,0) = \frac{N_{mn}}{ik}\pi w_0^2 \int_{-\infty}^{\infty}\int_{-\infty}^{\infty} dp_x dp_y \exp[-i2\pi(p_x x + p_y y)](-i2\pi p_x w_0)^m$$
$$\times (-i2\pi p_y w_0)^n \exp\left[-\pi^2 w_0^2\left(p_x^2 + p_y^2\right)\right]$$

(11)

From Eq. (11), the two-dimensional Fourier transform $\overline{a}_{x0}^{\pm}(p_x,p_y,0)$ of $a_{x0}^{\pm}(x,y,0)$ is recovered as

$$\overline{a}_{x0}^{\pm}(p_x,p_y,0) = \frac{N_{mn}}{ik}\pi w_0^2(-i2\pi p_x w_0)^m (-i2\pi p_y w_0)^n \exp\left[-\pi^2 w_0^2\left(p_x^2 + p_y^2\right)\right]$$

(12)

When Eq. (12) is substituted into Eq. (8), $\overline{a}_{x0}^{\pm}(p_x,p_y,z)$ is obtained as

$$\overline{a}_{x0}^{\pm}(p_x,p_y,z) = \frac{N_{mn}}{ik}\pi w_0^2(-i2\pi p_x w_0)^m (-i2\pi p_y w_0)^n \exp\left[-\frac{\pi^2 w_0^2(p_x^2 + p_y^2)}{q_{\pm}^2}\right]$$

(13)

where

$$q_{\pm} = \left(1 \pm \frac{iz}{b}\right)^{-1/2}$$

(14)

The subscript $\pm$ is also used to indicate that the propagation is in the $\pm z$ direction. The inverse Fourier transform of Eq. (13) is taken, with the result that

$$a_{x0}^{\pm}(x,y,z) = \frac{N_{mn}}{ik}\pi w_0^2 \int_{-\infty}^{\infty}\int_{-\infty}^{\infty} dp_x dp_y \exp[-i2\pi(p_x x + p_y y)](-i2\pi p_x w_0)^m$$
$$\times (-i2\pi p_y w_0)^n \exp\left[-\frac{\pi^2 w_0^2(p_x^2 + p_y^2)}{q_{\pm}^2}\right]$$

(15)

$$= \frac{N_{mn}}{ik}\pi w_0^2 w_0^{m+n} \frac{\partial^m}{\partial x^m}\frac{\partial^n}{\partial y^n}\int_{-\infty}^{\infty}\int_{-\infty}^{\infty} dp_x dp_y \exp[-i2\pi(p_x x + p_y y)]$$
$$\times \exp\left[-\frac{\pi^2 w_0^2(p_x^2 + p_y^2)}{q_{\pm}^2}\right]$$

(16)

For the position coordinates in the two physical spaces $0 < z < \infty$ and $-\infty < z < 0$, $1/q_{\pm}^2 \neq 0$, and the intergrals in Eq. (16) are evaluated, with the result that

$$a_{x0}^{\pm}(x,y,z) = \frac{N_{mn}}{ik} w_0^{m+n} \frac{\partial^m}{\partial x^m} \frac{\partial^n}{\partial y^n} q_{\pm}^2 \exp\left[-\frac{q_{\pm}^2(x^2+y^2)}{w_0^2}\right] \tag{17}$$

Using Eq. (2), Eq. (17) is expressed in terms of Hermite polynomials as follows:

$$a_{x0}^{\pm}(x,y,z) = \frac{N_{mn}}{ik} q_{\pm}^{m+n+2}(-1)^{m+n} H_m\left(\frac{q_{\pm}x}{w_0}\right) H_n\left(\frac{q_{\pm}y}{w_0}\right) \exp\left[-\frac{q_{\pm}^2(x^2+y^2)}{w_0^2}\right] \tag{18}$$

From Eqs. (5) and (18), the vector potential governing the linearly polarized electromagnetic complex-argument Hermite–Gauss beam is obtained as

$$A_{x0}^{\pm}(x,y,z) = \exp(\pm ikz) \frac{N_{mn}}{ik} q_{\pm}^{m+n+2}(-1)^{m+n}$$

$$\times H_m\left(\frac{q_{\pm}x}{w_0}\right) H_n\left(\frac{q_{\pm}y}{w_0}\right) \exp\left[-\frac{q_{\pm}^2(x^2+y^2)}{w_0^2}\right] \tag{19}$$

From Eq. (5), it follows that

$$\overline{A}_{x0}^{\pm}(p_x,p_y,z) = \exp(\pm ikz)\overline{a}_{x0}^{\pm}(p_x,p_y,z) \tag{20}$$

An important feature of the complex-argument Hermite–Gauss beam is that it has a closed-form Fourier transform representation, as given by Eqs. (13) and (20).

1.2 Time-averaged power

The electromagnetic fields associated with the paraxial beam are seen from Eq. (1.12) to be given by

$$E_{x0}^{\pm}(x,y,z) = \pm H_{y0}^{\pm}(x,y,z) = ikA_{x0}^{\pm}(x,y,z) \tag{21}$$

$A_{x0}^{\pm}(x,y,z)$ and $E_{x0}^{\pm}(x,y,z)$ are found from Eqs. (19) and (21) to be continuous, but $H_{y0}^{\pm}(x,y,z)$ is seen to be discontinuous across the secondary source plane $z = 0$. The discontinuity of $H_{y0}^{\pm}(x,y,z)$ is equivalent to an electric current density induced on the $z = 0$ plane. The induced electric current density is obtained from Eq. (21) as

$$\mathbf{J}_0(x,y,z) = \hat{z} \times \hat{y}[H_y^+(x,y,0) - H_y^-(x,y,0)]\delta(z) = -\hat{x}2ikA_{x0}^+(x,y,0)\delta(z) \tag{22}$$

By the use of Eqs. (5) and (22), the complex power is determined as

$$P_C = P_{re} + iP_{im} = -\frac{c}{2} \int_{-\infty}^{\infty} \int_{-\infty}^{\infty} \int_{-\infty}^{\infty} dxdydz\, \mathbf{E}_0^{\pm}(x,y,z) \cdot \mathbf{J}_0^{*}(x,y,z)$$

$$= ck^2 \int_{-\infty}^{\infty} \int_{-\infty}^{\infty} dxdya_{x0}^{\pm}(x,y,0)a_{x0}^{\pm *}(x,y,0) \tag{23}$$

P_C, given by Eq. (23), is real. Therefore, P_{im}, the imaginary part of P_C, is zero, yielding

$$P_{im} = 0 \tag{24}$$

The reactive power of the complex-argument Hermite–Gauss beam vanishes.

The Fourier transform representation of $a_{x0}^\pm(x, y, 0)$ is used in Eq. (23) to obtain

$$P_{re} = ck^2 \int_{-\infty}^{\infty} \int_{-\infty}^{\infty} dx dy \int_{-\infty}^{\infty} \int_{-\infty}^{\infty} dp_x dp_y \exp[-i2\pi(p_x x + p_y y)] a_{x0}^\pm(p_x, p_y, 0)$$

$$\times \int_{-\infty}^{\infty} \int_{-\infty}^{\infty} d\overline{p}_x d\overline{p}_y \exp[i2\pi(\overline{p}_x x + \overline{p}_y y)] \overline{a}_{x0}^{\pm*}(\overline{p}_x, \overline{p}_y, 0) \tag{25}$$

The integrations with respect to x and y are carried out to yield $\delta(p_x - \overline{p}_x)$ and $\delta(p_y - \overline{p}_y)$, respectively. The integrations with respect to $\overline{p}_x$ and $\overline{p}_y$ are performed. The result is

$$P_{re} = ck^2 \int_{-\infty}^{\infty} \int_{-\infty}^{\infty} dp_x dp_y \overline{a}_{x0}^\pm(p_x, p_y, 0) \overline{a}_{x0}^{\pm*}(p_x, p_y, 0) \tag{26}$$

Substituting Eq. (12) into Eq. (26) yields

$$P_{re} = c N_{mn}^2 \pi^2 w_0^4 (4\pi^2 w_0^2)^{m+n} I(m, w_0) I(n, w_0) \tag{27}$$

where

$$I(m, w_0) = \int_{-\infty}^{\infty} dp_x p_x^{2m} \exp\left(-2\pi^2 w_0^2 p_x^2\right) \tag{28}$$

and

$$I(n, w_0) = \int_{-\infty}^{\infty} dp_y p_y^{2n} \exp\left(-2\pi^2 w_0^2 p_y^2\right) \tag{29}$$

Substituting $2^{1/2}\pi w_0 p_x = \xi^{1/2}$, $I(m, w_0)$ is simplified as

$$I(m, w_0) = \frac{1}{2^{m+1/2}\pi^{2m+1}w_0^{2m+1}} \int_0^{\infty} d\xi \xi^{m+1/2-1} \exp(-\xi) \tag{30}$$

The integral in Eq. (30) is expressed as a Gamma function [6]. Hence, we have that

$$I(m, w_0) = \frac{\Gamma(m + 1/2)}{2^{m+1/2}\pi^{2m+1}w_0^{2m+1}} \tag{31}$$

When $I(m, w_0)$ and $I(n, w_0)$ are substituted into Eq. (27), P_{re} is obtained as

$$P_{re} = N_{mn}^2 c w_0^2 2^{m+n-1} \Gamma\left(m + \frac{1}{2}\right)\Gamma\left(n + \frac{1}{2}\right) \tag{32}$$

The normalization constant is chosen as

$$N_{mn} = \left[c w_0^2 2^{m+n-2} \Gamma\left(m + \frac{1}{2}\right)\Gamma\left(n + \frac{1}{2}\right)\right]^{-1/2} \tag{33}$$

with the result that

$$P_{re} = 2 \, \text{W} \tag{34}$$

The time-averaged power transported by the paraxial beam in the $\pm z$ direction is denoted by $P_{CH}^{\pm}$. By symmetry, $P_{CH}^{+} = P_{CH}^{-}$. The real power P_{re} is the time-averaged power generated by the current source. One half of this power is transported in the $+z$ direction and the other half is carried in the $-z$ direction. It follows from Eq. (34) that

$$P_{CH}^{+} = P_{CH}^{-} = \frac{1}{2} P_{re} = 1 \, \text{W} \tag{35}$$

The normalization constant N_{mn} is chosen such that the time-averaged power $P_{CH}^{\pm}$ transported by the paraxial beam in the $\pm z$ direction is given by $P_{CH}^{\pm} = 1$ W.

By the use of the relations

$$\Gamma(z+1) = z\Gamma(z) \tag{36}$$

and

$$\Gamma\left(\frac{1}{2}\right) = \pi^{1/2} \tag{37}$$

N_{mn} is transformed into a more convenient form as

$$N_{mn} = \left[\frac{4m!n!2^{m+n}}{c\pi w_0^2 (2m)!(2n)!} \right]^{1/2} \tag{38}$$

For $m = n = 0$, Eq. (19) reduces to the fundamental Gaussian beam given by Eq. (1.11). The normalization constant N_{mn}, given by Eq. (38), correctly reproduces the normalization constant N of the fundamental Gaussian beam, as given by Eq. (1.2).

2 Complex-argument Hermite–Gauss wave

To obtain the source in the complex space for the full-wave generalization of the complex-argument Hermite–Gauss beam, the search for the source is made at $|z| - ib = 0$. Then, from Eqs. (15) and (16), the complex space source is obtained in two different forms as

$$C_{s,mn}(x,y,z) = \frac{N_{mn}}{ik} \pi w_0^2 \int_{-\infty}^{\infty} \int_{-\infty}^{\infty} dp_x dp_y \exp[-i2\pi(p_x x + p_y y)]$$
$$\times (-i2\pi p_x w_0)^m (-i2\pi p_y w_0)^n \tag{39}$$

$$= \frac{N_{mn}}{ik} \pi w_0^2 w_0^{m+n} \frac{\partial^m}{\partial x^m} \delta(x) \frac{\partial^n}{\partial x^n} \delta(y) \tag{40}$$

The form of the source as given by Eq. (40) is necessary only to establish that the source is highly localized in the complex space. The complex space source in the form given by Eq. (39) is sufficient for deducing the expression for the full wave.

The source given by Eqs. (39) and (40) located at $|z| - ib = 0$ generates the paraxial beam for $|z| > 0$. The same source moved to $|z| = 0$ produces only the asymptotic limit of the paraxial beam. The paraxial equation is only an approximation for the Helmholtz equation. Therefore, the complex space source deduced from the paraxial beam, except for an excitation coefficient S_{ex}, is used for the Helmholtz equation to obtain the asymptotic limit of the full wave. Hence, the source that gives rise to the asymptotic limit of the full-wave solution of the Helmholtz equation is determined from Eq. (39) as

$$C_{s\infty,mn}(x,y,z) = S_{ex} \frac{N_{mn}}{ik} \pi w_0^2 \delta(z) \int_{-\infty}^{\infty} \int_{-\infty}^{\infty} dp_x dp_y \, \exp[-i2\pi(p_x x + p_y y)]$$
$$\times (-i2\pi p_x w_0)^m (-i2\pi p_y w_0)^n \tag{41}$$

Let $G_{m,n}(x,y,z)$ be the solution of the Helmholtz equation for the source given by Eq. (41). Then, $G_{m,n}(x,y,z)$ satisfies the following differential equation:

$$\left(\frac{\partial^2}{\partial x^2} + \frac{\partial^2}{\partial y^2} + \frac{\partial^2}{\partial z^2} + k^2\right) G_{m,n}(x,y,z) = -S_{ex} \frac{N_{mn}}{ik} \pi w_0^2 \delta(z) \int_{-\infty}^{\infty} \int_{-\infty}^{\infty} dp_x dp_y$$
$$\times \exp[-i2\pi(p_x x + p_y y)]$$
$$\times (-i2\pi p_x w_0)^m (-i2\pi p_y w_0)^n \tag{42}$$

$G_{m,n}(x,y,z)$ is expressed in terms of its two-dimensional Fourier transform $\overline{G}_{m,n}(p_x,p_y,z)$. From Eq. (42), the differential equation satisfied by $\overline{G}_{m,n}(p_x,p_y,z)$ is then determined as

$$\left(\frac{\partial^2}{\partial z^2} + \zeta^2\right) \overline{G}_{m,n}(p_x,p_y,z) = -S_{ex} \frac{N_{mn}}{ik} \pi w_0^2 (-i2\pi p_x w_0)^m (-i2\pi p_y w_0)^n \delta(z) \tag{43}$$

where

$$\zeta = \left[k^2 - 4\pi^2 \left(p_x^2 + p_y^2\right)\right]^{1/2} \tag{44}$$

The solution of Eq. (43) is found as

$$\overline{G}_{m,n}(p_x,p_y,z) = \frac{iS_{ex}}{2} \frac{N_{mn}}{ik} \pi w_0^2 (-i2\pi p_x w_0)^m (-i2\pi p_y w_0)^n \zeta^{-1} \exp(i\zeta|z|) \tag{45}$$

The inverse Fourier transform of Eq. (45) yields

$$G_{m,n}(x,y,z) = \frac{iS_{ex}}{2} \frac{N_{mn}}{ik} \pi w_0^2 \int_{-\infty}^{\infty} \int_{-\infty}^{\infty} dp_x dp_y \, \exp[-i2\pi(p_x x + p_y y)]$$
$$\times (-i2\pi p_x w_0)^m (-i2\pi p_y w_0)^n \zeta^{-1} \exp(i\zeta|z|) \tag{46}$$

By analytically continuing Eq. (46) from $|z|$ to $|z| - ib$, $A_x^{\pm}(x, y, z)$ is determined as

$$A_x^{\pm}(x, y, z) = \frac{iS_{ex}}{2} \frac{N_{mn}}{ik} \pi w_0^2 \int_{-\infty}^{\infty} \int_{-\infty}^{\infty} dp_x dp_y \, \exp[-i2\pi(p_x x + p_y y)]$$
$$\times (-i2\pi p_x w_0)^m (-i2\pi p_y w_0)^n \, \zeta^{-1} \exp[i\zeta(|z| - ib)] \qquad (47)$$

The paraxial approximation $[4\pi^2(p_x^2 + p_y^2)/k^2 \ll 1]$ of Eq. (47) is derived as

$$A_x^{\pm}(x, y, z) = \exp(\pm ikz) \frac{iS_{ex}}{2k} \exp(kb) \frac{N_{mn}}{ik} \pi w_0^2 \int_{-\infty}^{\infty} \int_{-\infty}^{\infty} dp_x dp_y$$
$$\times \exp[-i2\pi(p_x x + p_y y)] \, (-i2\pi p_x w_0)^m (-i2\pi p_y w_0)^n$$
$$\times \exp\left[-\frac{\pi^2 w_0^2 \left(p_x^2 + p_y^2\right)}{q_{\pm}^2}\right] \qquad (48)$$

A comparison of Eqs. (5) and (15) with Eq. (48) reveals that $A_x^{\pm}(x, y, z)$, given by Eq. (47), correctly reproduces the initially chosen paraxial beam if the excitation coefficient is chosen as

$$S_{ex} = -i2k \exp(-kb) \qquad (49)$$

When Eq. (49) is substituted into Eq. (47), $A_x^{\pm}(x, y, z)$ is found as

$$A_x^{\pm}(x, y, z) = k \exp(-kb) \frac{N_{mn}}{ik} \pi w_0^2 \int_{-\infty}^{\infty} \int_{-\infty}^{\infty} dp_x dp_y \, \exp[-i2\pi(p_x x + p_y y)]$$
$$\times (-i2\pi p_x w_0)^m (-i2\pi p_y w_0)^n \, \zeta^{-1} \exp[i\zeta(|z| - ib)] \qquad (50)$$

The full-wave generalization represented by Eq. (50) is designated the basic full complex-argument Hermite–Gauss wave.

3 Real and reactive powers

To obtain the complex power, only the field components $E_x^{\pm}(x, y, z)$ and $H_y^{\pm}(x, y, z)$ are needed. From Eqs. (2.10) and (50), $H_y^{\pm}(x, y, z)$ is found as

$$H_y^{\pm}(x, y, z) = \pm\exp(-kb)N_{mn}\pi w_0^2 \int_{-\infty}^{\infty} \int_{-\infty}^{\infty} dp_x dp_y \, \exp[-i2\pi(p_x x + p_y y)]$$
$$\times (-i2\pi p_x w_0)^m (-i2\pi p_y w_0)^n \, \exp[i\zeta(|z| - ib)] \qquad (51)$$

The electric current density induced on the $z = 0$ plane is determined from Eq. (51) as

$$\mathbf{J}(x,y,z) = \hat{z} \times \hat{y}[H_y^+(x,y,0) - H_y^-(x,y,0)]\delta(z)$$

$$= -\hat{x}2 \exp(-kb)N_{mn}\pi w_0^2\delta(z)\int_{-\infty}^{\infty}\int_{-\infty}^{\infty}d\overline{p}_x d\overline{p}_y$$

$$\times \exp[-i2\pi(\overline{p}_x x + \overline{p}_y y)] \, (-i2\pi\overline{p}_x w_0)^m(-i2\pi\overline{p}_y w_0)^n \exp(\overline{\zeta}b) \tag{52}$$

where $\overline{\zeta}$ is the same as ζ, with p_x and p_y replaced by $\overline{p}_x$ and $\overline{p}_y$, respectively. The complex power associated with the electric current is given by [see Eq. (D18)]

$$P_C = P_{re} + iP_{im} = -\frac{c}{2}\int_{-\infty}^{\infty}\int_{-\infty}^{\infty}\int_{-\infty}^{\infty}dx\,dy\,dz \, \mathbf{E}^{\pm}(x,y,z)\cdot\mathbf{J}^*(x,y,z) \tag{53}$$

The integration with respect to z is carried out. From Eqs. (52) and (53), P_C is obtained as

$$P_C = P_{re} + iP_{im} = -\frac{c}{2}\int\int dx\,dy\, E_x^{\pm}(x,y,0)J_x^*(x,y,0) \tag{54}$$

where $J_x(x,y,0)$ is the x component of $\mathbf{J}(x,y,z)$ divided by $\delta(z)$. From Eqs. (2.9) and (50), $E_x^{\pm}(x,y,z)$ is deduced as

$$E_x^{\pm}(x,y,z) = k \exp(-kb)N_{mn}\pi w_0^2\int_{-\infty}^{\infty}\int_{-\infty}^{\infty}dp_x dp_y$$

$$\times \exp[-i2\pi(p_x x + p_y y)]\left(1 - \frac{4\pi^2 p_x^2}{k^2}\right)(-i2\pi p_x w_0)^m(-i2\pi p_y w_0)^n$$

$$\times \zeta^{-1}\exp[i\zeta(|z| - ib)] \tag{55}$$

Substituting $E_x^{\pm}(x,y,z)$ from Eq. (55) and $J_x(x,y,0)$ from Eqs. (52) into Eq. (54) leads to the complex power as

$$P_C = -\frac{c}{2}\int_{-\infty}^{\infty}\int_{-\infty}^{\infty}dx\,dy\,k \exp(-kb)N_{mn}\pi w_0^2\int_{-\infty}^{\infty}\int_{-\infty}^{\infty}dp_x dp_y \exp[-i2\pi(p_x x + p_y y)]$$

$$\times \left(1 - \frac{4\pi^2 p_x^2}{k^2}\right)(-i2\pi p_x w_0)^m (-i2\pi p_y w_0)^n \zeta^{-1}\exp(\zeta b)$$

$$\times (-2)\exp(-kb)N_{mn}\pi w_0^2\int_{-\infty}^{\infty}\int_{-\infty}^{\infty}d\overline{p}_x d\overline{p}_y \exp[i2\pi(\overline{p}_x x + \overline{p}_y y)]$$

$$\times (i2\pi\overline{p}_x w_0)^m (i2\pi\overline{p}_y w_0)^n \exp(\overline{\zeta}^* b) \tag{56}$$

The integrations with respect to x and y are carried out first to yield $\delta(p_x - \bar{p}_x)$ and $\delta(p_y - \bar{p}_y)$, respectively. The integrations with respect to $\bar{p}_x$ and $\bar{p}_y$ are performed next. The result is

$$P_C = c\,\exp(-2kb)N_{mn}^2\pi^2 kw_0^4 \int_{-\infty}^{\infty}\int_{-\infty}^{\infty} dp_x dp_y \left(1 - \frac{4\pi^2 p_x^2}{k^2}\right)\left(4\pi^2 p_x^2 w_0^2\right)^m$$

$$\times \left(4\pi^2 p_y^2 w_0^2\right)^n \zeta^{-1}\exp[b(\zeta + \zeta^*)] \tag{57}$$

The integration variables are changed as

$$2\pi p_x = p\cos\phi \quad \text{and} \quad 2\pi p_y = p\sin\phi \tag{58}$$

Then, Eq. (57) simplifies as

$$P_C = \frac{cN_{mn}^2 w_0^2}{4}\exp(-2kb)w_0^{2(m+n+1)}\int_0^{\infty} dp\,p\,p^{2(m+n)}$$

$$\times \int_0^{2\pi} d\phi\,\cos^{2m}\phi\sin^{2n}\phi\left(1 - \frac{p^2\cos^2\phi}{k^2}\right)\xi^{-1}\exp[kb(\xi + \xi^*)] \tag{59}$$

where

$$\xi = \left(1 - \frac{p^2}{k^2}\right)^{1/2} \quad \text{for } 0 < p < k$$

$$= i\left(\frac{p^2}{k^2} - 1\right)^{1/2} \quad \text{for } k < p < \infty \tag{60}$$

The reactive power P_{im}, the imaginary part of P_C, arises from p in the range $k < p < \infty$, where ξ is imaginary. Hence, from Eqs. (59) and (60), P_{im} is determined as

$$P_{im} = -\frac{cN_{mn}^2 w_0^2}{4}\exp(-2kb)w_0^{2(m+n+1)}\int_k^{\infty} dp\,p\,p^{2(m+n)}\int_0^{2\pi} d\phi\,\cos^{2m}\phi\sin^{2n}\phi$$

$$\times \left(1 - \frac{p^2\cos^2\phi}{k^2}\right)\left(\frac{p^2}{k^2} - 1\right)^{-1/2} \tag{61}$$

Introducing a new variable of integration as given by $p^2 = k^2(1 + \tau^2)$, the integral in Eq. (61) is obtained as

$$I_{k\infty} = k^{2(m+n+1)}\int_0^{\infty} d\tau\left(1 + \tau^2\right)^{m+n}\int_0^{2\pi} d\phi\,\cos^{2m}\phi\sin^{2n}\phi[1 - (1 + \tau^2)\cos^2\phi] = \infty \tag{62}$$

The value of the integral is infinite as indicated in Eq. (62). Therefore,

$$P_{im} = \infty \tag{63}$$

The reactive power P_{im} associated with the basic full complex-argument Hermite–Gauss wave is infinite. This result is to be expected since the complex space source [Eq. (40)] causing the full wave is a higher-order point source. It is necessary to find in the complex space a source distribution, instead of a point source, to obtain finite values of the reactive power. One possible method is to use the procedure that was carried out for the fundamental Gaussian wave and introduce an analytical continuation different from $|z|$ to $|z| - ib = 0$.

The real power P_{re}, the real part of P_C, arises from p in the range $0 < p < k$, where ξ is real. Therefore, from Eqs. (59) and (60), P_{re} is determined as

$$P_{re} = \frac{w_0^{2(m+n+1)}}{2^{m+n}\Gamma(m+1/2)\Gamma(n+1/2)} \exp(-2kb) \int_0^k dp\, p\, p^{2(m+n)} \int_0^{2\pi} d\phi \cos^{2m}\phi \sin^{2n}\phi$$

$$\times \left(1 - \frac{p^2}{k^2}\cos^2\phi\right) \xi^{-1}\exp(2kb\xi) \tag{64}$$

The value of N_{mn} from Eq. (33) is substituted in obtaining Eq. (64). Consequently, the normalization of the wave amplitude is such that the real power in the corresponding paraxial beam is $P_{re} = 2$ W, as given by Eq. (34). The integration variable is changed as $p = k\sin\theta$. Then, Eq. (64) is modified as

$$P_{re} = \frac{(kw_0)^{2(m+n+1)}}{2^{m+n}\Gamma(m+1/2)\Gamma(n+1/2)} \int_0^{\pi/2} d\theta \sin\theta \sin^{2(m+n)}\theta \exp[-k^2w_0^2(1-\cos\theta)]$$

$$\times \int_0^{2\pi} d\phi \cos^{2m}\phi \sin^{2n}\phi(1 - \cos^2\phi\sin^2\theta) \tag{65}$$

For the special case of $m = 0$ and $n = 0$, Eq. (65) correctly reproduces the real power of the basic full Gaussian wave (see Chapter 4).

The real power depends on kw_0, m, and n. In Eq. (65), the ϕ integrations are performed analytically and the θ integration is carried out numerically. For the mode numbers $m = 1$ and $n = 1$, the two ϕ integrals that arise are evaluated as follows.

$$\int_0^{2\pi} d\phi \cos^2\phi\sin^2\phi = \frac{\pi}{4} \tag{66}$$

$$\int_0^{2\pi} d\phi \cos^4\phi\sin^2\phi = \frac{\pi}{8} \tag{67}$$

From Eqs. (65)–(67), P_{re} for $m = n = 1$ is determined as

$$P_{re} = \frac{(kw_0)^6}{4} \int_0^{\pi/2} d\theta \sin\theta \sin^4\theta$$

$$\times \left(1 - \frac{1}{2}\sin^2\theta\right) \exp[-k^2w_0^2(1-\cos\theta)] \quad \text{for } m = 1,\, n = 1 \tag{68}$$

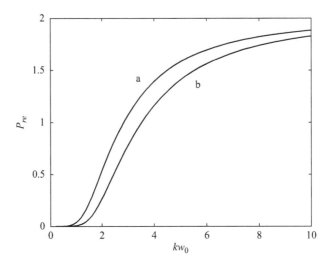

Fig. 1. Real power P_{re} as a function of kw_0 for $0.2 < kw_0 < 10$ and (a) $m = 1$, $n = 1$ and (b) $m = 1$, $n = 2$. The normalization is such that the real power in the corresponding paraxial beam is 2 W.

For the mode numbers $m = 1$ and $n = 2$, the two ϕ integrals that occur are obtained as follows:

$$\int_0^{2\pi} d\phi \cos^2\phi \sin^4\phi = \frac{\pi}{8} \tag{69}$$

$$\int_0^{2\pi} d\phi \cos^4\phi \sin^4\phi = \frac{3\pi}{64} \tag{70}$$

From Eqs. (65), (69), and (70), P_{re} for $m = 1$ and $n = 2$ is obtained as

$$P_{re} = \frac{(kw_0)^8}{24} \int_0^{\pi/2} d\theta \sin\theta \sin^6\theta$$

$$\times \left(1 - \frac{3}{8}\sin^2\theta\right) \exp[-k^2 w_0^2 (1 - \cos\theta)] \quad \text{for } m = 1, n = 2 \tag{71}$$

The numerical results of the real power P_{re} for the mode numbers (a) $m = 1$, $n = 1$ and (b) $m = 1$, $n = 2$ are shown in Fig. 1 as functions of kw_0 for $0.2 < kw_0 < 10$. In general, the real power increases, approaching the limiting value of the paraxial beam as kw_0 is increased. The increase of P_{re} with the increase in kw_0 is monotonic. For sufficiently large and fixed kw_0, the real power decreases as the order of the mode increases.

References

1. A. E. Siegman, "Hermite-Gaussian functions of complex argument as optical beam eigen-functions," *J. Opt. Soc. Am.* **63**, 1093–1094 (1973).

2. A. E. Siegman, *Lasers* (University Science, Mill Valley, CA, 1986).

3. S. R. Seshadri, "Virtual source for a Hermite-Gauss beam," *Opt. Lett.* **28**, 595–597 (2003).

4. I. S. Gradshteyn and I. M. Ryzhik, *Table of integrals, Series, and Products* (Academic Press, New York, 1965).

5. M. Abramowitz and I. A. Stegun, *Handbook of Mathematical Functions* (U.S. Government Printing Office, Washington, DC, 1965).

6. W. Magnus and F. Oberhettinger, *Functions of Mathematical Physics* (Chelsea Publishing Company, New York, 1954), p. 1.

CHAPTER 12

Basic full real-argument Hermite–Gauss wave

Kogelnik and Li [1] and Marcuse [2] introduced the real-argument Hermite–Gauss beams in connection with laser beams and resonators. The real-argument Hermite–Gauss beams are another series of higher-order solutions to the paraxial wave equation in the Cartesian coordinate system. This series of eigenfunctions forms an orthonormal set. The fundamental Gaussian beam is the lowest-order, $(m, n) = (0, 0)$, mode in this set, where m and n are the mode numbers in the x and y directions, respectively. In this chapter, the real-argument Hermite–Gauss beams are treated. As for the other paraxial beams, the reactive power of the real-argument Hermite–Gauss beams vanishes. The source in the complex space required for the full-wave generalization of the real-argument Hermite–Gauss beams is obtained. The basic full real-argument Hermite–Gauss wave produced by the complex space source is derived. The real and the reactive powers of the basic full real-argument Hermite–Gauss waves are evaluated. As for the other basic full Gaussian waves, the reactive power is infinite. The characteristics of the real power are discussed. The real power increases as kw_0 is increased, reaching the limiting value of the paraxial beam. For a fixed kw_0, the real power decreases as the mode orders increase.

1 Real-argument Hermite–Gauss beam

1.1 Paraxial beam

For the excitation of a linearly polarized electromagnetic real-argument Hermite–Gauss beam, the required x component of the magnetic vector potential at the secondary source plane is specified as

$$a_{x0}^{\pm}(x, y, 0) = A_{x0}^{\pm}(x, y, 0) = \frac{N_{mn}}{ik} H_m\left(\frac{\sqrt{2}x}{w_0}\right) H_n\left(\frac{\sqrt{2}y}{w_0}\right) \exp\left(-\frac{x^2 + y^2}{w_0^2}\right) \tag{1}$$

The paraxial approximation of the x component of the magnetic vector potential is given by $A_{x0}^{\pm}(x, y, z)$. The slowly varying amplitude $a_{x0}^{\pm}(x, y, z)$ is related to $A_{x0}^{\pm}(x, y, z)$, as stated in

Eq. (11.5). The paraxial wave equation satisfied by $a_{x0}^\pm(x,y,z)$ and the differential equation satisfied by the Fourier transform $\bar{a}_{x0}^\pm(p_x,p_y,z)$ of $a_{x0}^\pm(x,y,z)$ are as expressed in Eqs. (11.6) and (11.7), respectively. The solution of $\bar{a}_{x0}^\pm(p_x,p_y,z)$ in terms of $\bar{a}_{x0}^\pm(p_x,p_y,0)$ as obtained in Eq. (11.8) is reproduced here as

$$\bar{a}_{x0}^\pm(p_x,p_y,z) = \bar{a}_{x0}^\pm(p_x,p_y,0)\exp\left[-\pi^2 w_0^2\left(p_x^2 + p_y^2\right)\frac{i|z|}{b}\right] \tag{2}$$

Consider the following Fourier transform:

$$I_m(p_x) = \int_{-\infty}^{\infty} dx\, \exp\left(i2\pi p_x x\right) H_m\left(\frac{\sqrt{2}x}{w_0}\right)\exp\left(-\frac{x^2}{w_0^2}\right) \tag{3}$$

The integral in Eq. (3) is evaluated by the use of the formula 7.376.1 in [3], with the result that

$$I_m(p_x) = \sqrt{\pi}w_0 i^m H_m\left(\sqrt{2}\pi w_0 p_x\right)\exp\left(-\pi^2 w_0^2 p_x^2\right) \tag{4}$$

In the two-dimensional Fourier transform of Eq. (1), two one-dimensional integrals of the form given by Eq. (3) occur. Therefore, using Eq. (4), the two-dimensional Fourier transform of Eq. (1) is obtained as

$$\bar{a}_{x0}^\pm(p_x,p_y,0) = \bar{A}_{x0}^\pm(p_x,p_y,0) = \frac{N_{mn}}{ik}\pi w_0^2 i^{m+n}$$
$$\times H_m\left(\sqrt{2}\pi w_0 p_x\right)H_n\left(\sqrt{2}\pi w_0 p_y\right)\exp\left[-\pi^2 w_0^2\left(p_x^2 + p_y^2\right)\right] \tag{5}$$

When Eq. (5) is substituted into Eq. (2), $\bar{a}_{x0}^\pm(p_x,p_y,z)$ is found as

$$\bar{a}_{x0}^\pm(p_x,p_y,z) = \frac{N_{mn}}{ik}\pi w_0^2 i^{m+n} H_m\left(\sqrt{2}\pi w_0 p_x\right)$$
$$\times H_n\left(\sqrt{2}\pi w_0 p_y\right)\exp\left[-\pi^2 w_0^2\frac{\left(p_x^2 + p_y^2\right)}{q_\pm^2}\right] \tag{6}$$

where

$$q_\pm = \left(1 \pm \frac{iz}{b}\right)^{-1/2} = q_\mp^* \tag{7}$$

As before, the sign $\pm$, the superscript $\pm$, and the subscript $\pm$ indicate that the propagation is in the $\pm z$ direction.

The inverse Fourier transform of Eq. (6) is taken, with the result that

$$a_{x0}^\pm(x,y,z) = \frac{N_{mn}}{ik}I_m(x,z)I_n(y,z) \tag{8}$$

where

$$I_m(x,z) = \pi^{1/2} w_0 i^m \int_{-\infty}^{\infty} dp_x \exp\left(-i2\pi p_x x\right) H_m\left(\sqrt{2}\pi w_0 p_x\right) \exp\left[-\pi^2 w_0^2 \frac{p_x^2}{q_\pm^2}\right] \tag{9}$$

and $I_n(y,z)$ is obtained from $I_m(x,z)$ by appropriate substitution. In Appendix AA, the method of evaluation of the integral $I_m(x,z)$ is presented. When $I_m(x,z)$ and $I_n(y,z)$, obtained from Eq. (AA9), are substituted into Eq. (8), $a_{x0}^\pm(x,y,z)$ is found as

$$a_{x0}^\pm(x,y,z) = \frac{N_{mn}}{ik} \frac{q_\pm^{m+n+2}}{q_\pm^{*(m+n)}} H_m\left(\frac{\sqrt{2}q_\pm q_\pm^* x}{w_0}\right) H_n\left(\frac{\sqrt{2}q_\pm q_\pm^* y}{w_0}\right) \exp\left[-\frac{q_\pm^2(x^2+y^2)}{w_0^2}\right] \tag{10}$$

The rapidly varying phase factor $\exp\left(\pm ikz\right)$ is included, and from Eq. (10), the vector potential governing the linearly polarized real-argument Hermite–Gauss beam is determined as

$$A_{x0}^\pm(x,y,z) = \exp\left(\pm ikz\right) \frac{N_{mn}}{ik} \frac{q_\pm^{m+n+2}}{q_\pm^{*(m+n)}} H_m\left(\frac{\sqrt{2}q_\pm q_\pm^* x}{w_0}\right)$$

$$\times H_n\left(\frac{\sqrt{2}q_\pm q_\pm^* y}{w_0}\right) \exp\left[-\frac{q_\pm^2(x^2+y^2)}{w_0^2}\right] \tag{11}$$

It should be noted that $A_{x0}^\pm(x,y,z)$ given by Eq. (11) is valid only for the position coordinates in the two physical spaces $0 < z < \infty$ and $-\infty < z < 0$. From Eq. (11.5), it follows that

$$\overline{A}_{x0}^\pm(p_x,p_y,z) = \exp\left(\pm ikz\right)\overline{a}_{x0}^\pm(p_x,p_y,z) \tag{12}$$

As for the complex-argument Hermite–Gauss beam, the real-argument Hermite–Gauss beam has a closed-form Fourier transform representation, as given by Eqs. (6) and (12).

1.2 Time-averaged power

The electromagnetic fields, their characteristics, the induced electric current density on the secondary source plane $z = 0$, and the complex power associated with the real-argument Hermite–Gauss beam are the same as those for the complex-argument Hermite–Gauss beam. $A_{x0}^\pm(x,y,z)$ and $a_{x0}^\pm(x,y,z)$ pertaining to the real-argument Hermite–Gauss beam have to be used. As for the complex-argument Hermite–Gauss beam, for the real-argument Hermite–Gauss beam the reactive power vanishes; that is,

$$P_{im} = 0 \tag{13}$$

Substituting $a_{x0}^\pm(x,y,z)$ from Eq. (1) into Eq. (11.23), the real power P_{re} is found as

$$P_{re} = cN_{mn}^2 \int_{-\infty}^{\infty} dx H_m^2\left(\frac{\sqrt{2}x}{w_0}\right) \exp\left(-\frac{2x^2}{w_0^2}\right) \int_{-\infty}^{\infty} dy H_n^2\left(\frac{\sqrt{2}y}{w_0}\right) \exp\left(-\frac{2y^2}{w_0^2}\right) \tag{14}$$

The eigenfunctions associated with the real-argument Hermite–Gauss beams satisfy the orthogonality relation:

$$\int_{-\infty}^{\infty} dx H_m(ax) H_\ell(ax) \exp\left(-a^2 x^2\right) = \frac{2^m m! \pi^{1/2}}{a} \delta_{m\ell} \tag{15}$$

for a real, where $\delta_{m\ell} = 0$ for $m \neq \ell$ and $\delta_{m\ell} = 1$ for $m = \ell$. See formula 7.374.1 in [3]. Applying the orthogonality relation given by Eq. (15) to Eq. (14), the real power is determined as

$$P_{re} = c\pi w_0^2 2^{m+n-1} m! n! N_{mn}^2 \tag{16}$$

The normalization constant is chosen as

$$N_{mn} = \left(\frac{4}{c\pi w_0^2 2^{m+n} m! n!}\right)^{1/2} \tag{17}$$

with the result that

$$P_{re} = 2 \text{ W} \tag{18}$$

The time-averaged power transported by the paraxial beam in the $\pm z$ direction is denoted by $P_{RH}^\pm$. By symmetry, $P_{RH}^+ = P_{RH}^-$. The real power P_{re} is the time-averaged power generated by the current source. One half of this power is transported in the $+z$ direction and the other half is carried in the $-z$ direction. It follows from Eq. (18) that

$$P_{RH}^+ = P_{RH}^- = \frac{1}{2} P_{re} = 1 \text{ W} \tag{19}$$

The normalization constant N_{mn} is chosen such that the time-averaged power $P_{RH}^\pm$ transported by the paraxial beam in the $\pm z$ direction is given by $P_{RH}^\pm = 1$ W. For $m = n = 0$, Eq. (11) reduces to the fundamental Gaussian beam, as given by Eq. (1.11). The normalization constant N_{mn} given by Eq. (17) reduces correctly to the normalization constant N of the fundamental Gaussian beam, as given by Eq. (1.2).

2 Real-argument Hermite–Gauss wave

The search for the complex space source for the full-wave generalization of the real-argument Hermite–Gauss beam is made at $|z| - ib = 0$. Then, $1/q_\pm^2 = 0$, and from Eqs. (8) and (9), the complex space source is determined as

$$C_{s,mn}(x, y, z) = \frac{N_{mn}}{ik} \pi w_0^2 i^{m+n} \int_{-\infty}^{\infty} \int_{-\infty}^{\infty} dp_x dp_y$$
$$\times \exp\left[-i2\pi(p_x x + p_y y)\right] H_m(\sqrt{2}\pi w_0 p_x) H_n(\sqrt{2}\pi w_0 p_y) \tag{20}$$

The complex space source in the form given by Eq. (20) is sufficient for deducing the expression for the full wave.

The source given by Eq. (20) located at $|z| - ib = 0$ generates the paraxial beam for $|z| > 0$. The same source placed at $|z| = 0$ causes only the asymptotic limit of the paraxial beam. The paraxial equation is only an approximation for the Helmholtz equation. Consequently, the complex space source derived from the paraxial beam, except for an excitation coefficient S_{ex}, is used as the source for the Helmholtz equation to determine the asymptotic limit of the full wave. Therefore, the source that produces the asymptotic limit of the full-wave solution of the Helmholtz equation is determined from Eq. (20) as

$$
C_{s\infty,mn}(x,y,z) = S_{ex} \frac{N_{mn}}{ik} \pi w_0^2 i^{m+n} \delta(z) \int_{-\infty}^{\infty} \int_{-\infty}^{\infty} dp_x dp_y
$$
$$
\times \exp\left[-i2\pi(p_x x + p_y y)\right] H_m\left(\sqrt{2}\pi w_0 p_x\right) H_n\left(\sqrt{2}\pi w_0 p_y\right) \tag{21}
$$

Let $G_{m,n}(x,y,z)$ be the solution of the Helmholtz equation for the source given by Eq. (21). Then, $G_{m,n}(x,y,z)$ satisfies the following differential equation:

$$
\left(\frac{\partial^2}{\partial x^2} + \frac{\partial^2}{\partial y^2} + \frac{\partial^2}{\partial z^2} + k^2\right) G_{m,n}(x,y,z) = -S_{ex} \frac{N_{mn}}{ik} \pi w_0^2 i^{m+n} \delta(z) \int_{-\infty}^{\infty} \int_{-\infty}^{\infty} dp_x dp_y
$$
$$
\times \exp\left[-i2\pi(p_x x + p_y y)\right]
$$
$$
\times H_m\left(\sqrt{2}\pi w_0 p_x\right) H_n\left(\sqrt{2}\pi w_0 p_y\right) \tag{22}
$$

From Eq. (22), the differential equation satisfied by the two-dimensional Fourier transform $\overline{G}_{m,n}(p_x, p_y, z)$ of $G_{m,n}(x,y,z)$ is obtained as

$$
\left(\frac{\partial^2}{\partial z^2} + \zeta^2\right) \overline{G}_{m,n}(p_x, p_y, z) = -S_{ex} \frac{N_{mn}}{ik} \pi w_0^2 i^{m+n}
$$
$$
\times H_m\left(\sqrt{2}\pi w_0 p_x\right) H_n\left(\sqrt{2}\pi w_0 p_y\right) \delta(z) \tag{23}
$$

where ζ is the same as that given by Eq. (11.44). The solution of Eq. (23) is found as

$$
\overline{G}_{m,n}(p_x, p_y, z) = \frac{iS_{ex}}{2} \frac{N_{mn}}{ik} \pi w_0^2 i^{m+n} H_m\left(\sqrt{2}\pi w_0 p_x\right) H_n\left(\sqrt{2}\pi w_0 p_y\right) \zeta^{-1} \exp\left(i\zeta|z|\right) \tag{24}
$$

$G_{m,n}(x,y,z)$ is found from Eq. (24) by the inverse Fourier transformation. By analytically continuing $G_{m,n}(x,y,z)$ from $|z|$ to $|z| - ib$, $A_x^{\pm}(x,y,z)$ is deduced. The paraxial approximation $A_{x0}^{\pm}(x,y,z)$ of $A_x^{\pm}(x,y,z)$ is then derived. When $A_{x0}^{\pm}(x,y,z)$ is compared with Eqs. (8) and (9) after the inclusion of the rapidly varying phase factor $\exp(\pm ikz)$, it is found that $A_x^{\pm}(x,y,z)$ reduces properly to the initially chosen paraxial beam if S_{ex} is chosen to be the same as that given by Eq. (11.49). This value of S_{ex} is substituted into the expression for $A_x^{\pm}(x,y,z)$. The final expression for $A_x^{\pm}(x,y,z)$ is given by

$$
A_x^{\pm}(x,y,z) = k \exp(-kb) \frac{N_{mn}}{ik} \pi w_0^2 i^{m+n} \int_{-\infty}^{\infty} \int_{-\infty}^{\infty} dp_x dp_y
$$
$$
\times \exp\left[-i2\pi(p_x x + p_y y)\right] H_m\left(\sqrt{2}\pi w_0 p_x\right)
$$
$$
\times H_n\left(\sqrt{2}\pi w_0 p_y\right) \zeta^{-1} \exp\left[i\zeta(|z| - ib)\right] \tag{25}
$$

The full-wave generalization represented by Eq. (25) is the basic full real-argument Hermite–Gauss wave.

3 Real and reactive powers

A comparison of Eq. (11.50) with Eq. (25) shows that by changing $(-i2\pi p_x w_0)^m(-i2\pi p_y w_0)^n$ to $i^{m+n}H_m(\sqrt{2}\pi w_0 p_x)H_n(\sqrt{2}\pi w_0 p_y)$, the vector potential, the electromagnetic fields, and the source current density of the complex-argument Hermite–Gauss wave are transformed to the corresponding quantities pertaining to the real-argument Hermite–Gauss wave. Therefore, from Eq. (11.56), the expression for the complex power of the real-argument Hermite–Gauss wave is obtained as

$$
\begin{aligned}
P_C = -\frac{c}{2}\int_{-\infty}^{\infty}\int_{-\infty}^{\infty} &\, dxdy\, k\exp(-kb)N_{mn}\pi w_0^2 \\
&\times i^{m+n}\int_{-\infty}^{\infty}\int_{-\infty}^{\infty} dp_x dp_y\, \exp[-i2\pi(p_x x + p_y y)] \\
&\times \left(1-\frac{4\pi^2 p_x^2}{k^2}\right)H_m(\sqrt{2}\pi w_0 p_x)H_n(\sqrt{2}\pi w_0 p_y)\zeta^{-1}\exp(\zeta b) \\
&\times (-2)\exp(-kb)N_{mn}\pi w_0^2(-i)^{m+n}\int_{-\infty}^{\infty}\int_{-\infty}^{\infty} d\bar{p}_x d\bar{p}_y\, \exp[i2\pi(\bar{p}_x x + \bar{p}_y y)] \\
&\times H_m(\sqrt{2}\pi w_0 \bar{p}_x)H_n\left(\sqrt{2}\pi w_0 \bar{p}_y\right)\exp(\zeta^* b)
\end{aligned}
\tag{26}
$$

The integrations with respect to x and y are carried out first; the integrations with respect to $\bar{p}_x$ and $\bar{p}_y$ are performed next. The result is

$$
\begin{aligned}
P_C = c\exp(-2kb)N_{mn}^2 \pi^2 k w_0^4 \int_{-\infty}^{\infty}\int_{-\infty}^{\infty} &\, dp_x dp_y \\
&\times \left(1-\frac{4\pi^2 p_x^2}{k^2}\right)H_m^2(\sqrt{2}\pi w_0 p_x)H_n^2(\sqrt{2}\pi w_0 p_y)\zeta^{-1}\exp[b(\zeta+\zeta^*)]
\end{aligned}
\tag{27}
$$

The integration variables are changed as $2\pi p_x = p\cos\phi$ and $2\pi p_y = p\sin\phi$. Then, Eq. (27) is transformed as

$$
\begin{aligned}
P_C = \frac{cN_{mn}^2 w_0^4}{4}\exp(-2kb)\int_0^{\infty} dp\, p\int_0^{2\pi} &\, d\phi\left(1-\frac{p^2\cos^2\phi}{k^2}\right)H_m^2\left(\frac{w_0 p\cos\phi}{\sqrt{2}}\right) \\
&\times H_n^2\left(\frac{w_0 p\sin\phi}{\sqrt{2}}\right)\xi^{-1}\exp[kb(\xi+\xi^*)]
\end{aligned}
\tag{28}
$$

where ξ is the same as that given by Eq. (11.60).

The reactive power P_{im}, the imaginary part of P_C, arises from p in the range $k < p < \infty$, where ξ is imaginary. Hence, from Eqs. (28) and (11.60), P_{im} is determined as

$$
\begin{aligned}
P_{im} = -\frac{w_0^2 \exp(-2kb)}{\pi 2^{m+n}m!n!}\int_k^{\infty} dp\, p\int_0^{2\pi} &\, d\phi\left(1-\frac{p^2\cos^2\phi}{k^2}\right)H_m^2\left(\frac{w_0 p\cos\phi}{\sqrt{2}}\right) \\
&\times H_n^2\left(\frac{w_0 p\sin\phi}{\sqrt{2}}\right)\left(\frac{p^2}{k^2}-1\right)^{-1/2}
\end{aligned}
\tag{29}
$$

The value of N_{mn} given by Eq. (17) for the real-argument Hermite–Gauss beam is substituted in obtaining Eq. (29). The reactive power P_{im} given by Eq. (29) is infinite. This result is proved as follows. The Hermite functions are replaced by their polynomial representation, as given by Eq. (11.3). The leading term in this expansion is $H_p(v) = 2^p v^p$. If only the leading terms are retained in $H_m^2(v)$ and $H_n^2(v)$, Eq. (29) is transformed as

$$P_{im} = -\frac{\exp(-k^2 w_0^2) w_0^{2(m+n+1)}}{\pi m! n!} \int_k^\infty dp p^{2(m+n)} \int_0^{2\pi} d\phi \cos^{2m}\phi \sin^{2n}\phi$$
$$\times \left(1 - \frac{p^2 \cos^2\phi}{k^2}\right)\left(\frac{p^2}{k^2} - 1\right)^{-1/2} \tag{30}$$

Let

$$\int_0^{2\pi} d\phi \cos^{2m}\phi \sin^{2n}\phi = \pi a_1 \tag{31}$$

and

$$\int_0^{2\pi} d\phi \cos^{2(m+1)}\phi \sin^{2n}\phi = \pi a_2 \tag{32}$$

Then, Eq. (30) simplifies as

$$P_{im} = -\frac{\exp(-k^2 w_0^2) w_0^{2(m+n+1)}}{m! n!} \int_k^\infty dp p^{2(m+n)} \left(a_1 - \frac{p^2}{k^2} a_2\right)\left(\frac{p^2}{k^2} - 1\right)^{-1/2} \tag{33}$$

The variable of integration is changed as $p^2 = k^2(1 + \tau^2)$, with the result that

$$P_{im} = -\frac{\exp(-k^2 w_0^2)(kw_0)^{2(m+n+1)}}{m! n!} \int_0^\infty d\tau (1 + \tau^2)^{m+n}\left[a_1 - a_2(1 + \tau^2)\right] = \infty \tag{34}$$

The value of the integral is infinite as indicated in Eq. (34). When all the terms in the polynomial representation of the Hermite functions are retained, there are a number of terms in the expression for P_{im}; each term is of the same form as in Eq. (34). If m is even, for every term m is even and varies from 0 to m. If m is odd, for every term m is odd and varies from 1 to m. The variation of n also satisfies this requirement. Thus, every term in the general expression for P_{im} becomes infinite. Therefore, the reactive power P_{im} associated with the basic full real-argument Hermite–Gauss wave is infinite. This result is to be expected since the complex space source causing the full wave is a finite series of higher-order point sources (see Appendix AB). It is necessary for the source in the complex space to be a series of current distributions of finite width to yield finite values of the reactive power.

The real power P_{re}, the real part of P_C, arises from p in the range $0 < p < k$, where ξ is real. Therefore, from Eqs. (17), (28), and (11.60), P_{re} is obtained as

$$P_{re} = \frac{w_0^2}{\pi 2^{m+n} m! n!} \int_0^k dp p \int_0^{2\pi} d\phi \left(1 - \frac{p^2 \cos^2\phi}{k^2}\right) H_m^2\left(\frac{w_0 p \cos\phi}{\sqrt{2}}\right) H_n^2\left(\frac{w_0 p \sin\phi}{\sqrt{2}}\right)$$
$$\times \xi^{-1} \exp\left[-k^2 w_0^2(1 - \xi)\right] \tag{35}$$

The use of N_{mn} from Eq. (17) implies that the normalization of the wave amplitude is such that the real power in the corresponding paraxial beam is $P_{re} = 2$ W, as given by Eq. (18). The integration variable is changed as $p = k\sin\theta$. Then P_{re}, given by Eq. (35), is found as

$$P_{re} = \frac{k^2 w_0^2}{\pi 2^{m+n} m! n!} \int_0^{\pi/2} d\theta \sin\theta \int_0^{2\pi} d\phi (1 - \cos^2\phi \sin^2\theta)$$

$$\times H_m^2 \left(\frac{w_0 k \sin\theta \cos\phi}{\sqrt{2}}\right) H_n^2 \left(\frac{w_0 k \sin\theta \sin\phi}{\sqrt{2}}\right) \exp\left[-k^2 w_0^2 (1 - \cos\theta)\right] \qquad (36)$$

The Hermite functions are replaced by their polynomial representations, as given by Eq. (11.3). The ϕ integrations are carried out analytically. The θ integration is performed numerically. Then, the real power P_{re} is obtained as a function of kw_0, m, and n.

For some lower-order mode numbers (m, n), the expressions for P_{re} as integrals with respect to θ are as follows:

$$P_{re} = 2k^2 w_0^2 \int_0^{\pi/2} d\theta \, \sin\theta \left(1 - \frac{1}{2}\sin^2\theta\right)$$

$$\times \exp\left[-k^2 w_0^2 (1 - \cos\theta)\right] \quad \text{for } (m, n) = (0, 0) \qquad (37)$$

$$P_{re} = k^4 w_0^4 \int_0^{\pi/2} d\theta \, \sin\theta \sin^2\theta \left(1 - \frac{3}{4}\sin^2\theta\right)$$

$$\times \exp\left[-k^2 w_0^2 (1 - \cos\theta)\right] \quad \text{for } (m, n) = (1, 0) \qquad (38)$$

$$P_{re} = k^4 w_0^4 \int_0^{\pi/2} d\theta \, \sin\theta \sin^2\theta \left(1 - \frac{1}{4}\sin^2\theta\right)$$

$$\times \exp\left[-k^2 w_0^2 (1 - \cos\theta)\right] \quad \text{for } (m, n) = (0, 1) \qquad (39)$$

$$P_{re} = \frac{k^6 w_0^6}{4} \int_0^{\pi/2} d\theta \, \sin\theta \sin^4\theta \left(1 - \frac{1}{2}\sin^2\theta\right)$$

$$\times \exp\left[-k^2 w_0^2 (1 - \cos\theta)\right] \quad \text{for } (m, n) = (1, 1) \qquad (40)$$

$$P_{re} = \frac{k^2 w_0^2}{2} \int_0^{\pi/2} d\theta \, \sin\theta \left[k^4 w_0^4 \left(\frac{3}{4}\sin^4\theta - \frac{5}{8}\sin^6\theta\right)\right.$$

$$\left. - k^2 w_0^2 \left(2\sin^2\theta - \frac{3}{2}\sin^4\theta\right) + 2 - \sin^2\theta\right]$$

$$\times \exp\left[-k^2 w_0^2 (1 - \cos\theta)\right] \quad \text{for } (m, n) = (2, 0) \qquad (41)$$

$$P_{re} = \frac{k^2 w_0^2}{2} \int_0^{\pi/2} d\theta \, \sin\theta \left[k^4 w_0^4 \left(\frac{3}{4}\sin^4\theta - \frac{1}{8}\sin^6\theta \right) \right.$$
$$\left. - k^2 w_0^2 \left(2\sin^2\theta - \frac{1}{2}\sin^4\theta \right) + 2 - \sin^2\theta \right]$$
$$\times \exp\left[-k^2 w_0^2 (1 - \cos\theta)\right] \quad \text{for } (m,n) = (0,2) \tag{42}$$

Some details for deducing the expression of P_{re} for the mode numbers $(m,n) = (0,2)$ are as follows. From Eq. (11.3), it is seen that $H_0(v) = 1$ and $H_2(v) = 4v^2 - 2$. Therefore, we obtain that

$$H_2^2 \left(\frac{w_0 k \sin\theta \sin\phi}{\sqrt{2}} \right) = 4(w_0^4 k^4 \sin^4\theta \sin^4\phi - 2w_0^2 k^2 \sin^2\theta \sin^2\phi + 1) \tag{43}$$

Substituting for (m,n), $H_0^2(\ )$, and $H_2^2(\ )$ in Eq. (36) yields

$$P_{re} = \frac{k^2 w_0^2}{2\pi} \int_0^{\pi/2} d\theta \sin\theta \int_0^{2\pi} d\phi \left[w_0^4 k^4 (\sin^4\theta \sin^4\phi - \sin^6\theta \sin^4\phi \cos^2\phi) \right.$$
$$\left. - 2w_0^2 k^2 (\sin^2\theta \sin^2\phi - \sin^4\theta \sin^2\phi \cos^2\phi) + 1 - \sin^2\theta \cos^2\phi \right] \tag{44}$$

When the ϕ integrals are carried out, the result given by Eq. (42) is obtained.

The numerical results of the real power P_{re} are shown in Figs. 1–3 as functions of kw_0 for $0.2 < kw_0 < 10$. In Fig. 1, the results for the mode numbers $(m,n) = (0,0)$, $(1,0)$, and

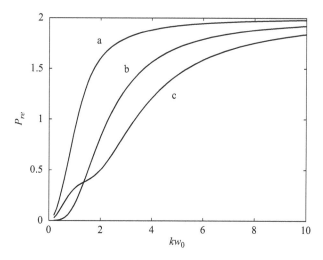

Fig. 1. Real power P_{re} as a function of kw_0 for $0.2 < kw_0 < 10$, and (a) $(m,n) = (0,0)$, (b) $(m,n) = (1,0)$, and (c) $(m,n) = (2,0)$. The normalization is such that the real power in the corresponding paraxial beam is 2 W.

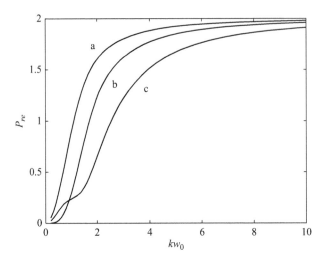

Fig. 2.   Same as in Fig. 1, but for (a) $(m, n) = (0, 0)$, (b) $(m, n) = (0, 1)$, and (c) $(m, n) = (0, 2)$.

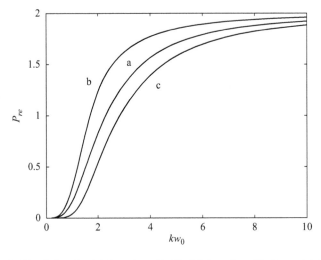

Fig. 3.   Same as in Fig. 1, but for (a) $(m, n) = (1, 0)$, (b) $(m, n) = (0, 1)$, and (c) $(m, n) = (1, 1)$.

$(2, 0)$ are shown. Figure 2 has the results for $(m, n) = (0, 0)$, $(0, 1)$, and $(0, 2)$. The results for $(m, n) = (1, 0)$, $(0, 1)$, and $(1, 1)$ are contained in Fig. 3. In general, the real power increases, approaching the limiting value of the paraxial beam as kw_0 is increased. The increase of P_{re} with the increase in kw_0 is monotonic. In Fig. 1, $n = 0$, but m increases from 0 to 2. For sufficiently large and fixed kw_0, P_{re} decreases as the mode order m increases. In Fig. 2, $m = 0$, but n increases from 0 to 2. In Fig. 2, as in Fig. 1, for large and fixed kw_0, P_{re} decreases as the mode order n increases. In Fig. 3, comparing the results for the mode numbers $(1, 0)$ and $(1, 1)$, for fixed kw_0, P_{re} decreases as n increases; similarly, comparing the results for the mode numbers $(0, 1)$ and $(1, 1)$, again for fixed kw_0, P_{re} decreases as m increases. Thus, in general, for sufficiently large and fixed kw_0, P_{re} decreases as the mode order m or n increases.

For the special case of $(m, n) = (0, 0)$, Eq. (36) reduces correctly to the real power of the basic full Gaussian wave (see Chapter 4). For $(m, n) = (1, 0)$, $(0, 1)$, and $(1, 1)$, the results given by Eqs. (38), (39), and (40), respectively, for the basic full real-argument Hermite–Gauss wave are identical to the corresponding results of the basic full complex-argument Hermite–Gauss wave, as given by Eq. (11.65). Therefore, the results depicted in Fig. 3 apply also to the basic full complex-argument Hermite–Gauss wave. Thus, the results for $m > 1$ or/and $n > 1$ are valid only for the basic full real-argument Hermite–Gauss wave.

Appendix AA: Evaluation of the integral $I_m(x,z)$

The generating function for $H_m(\xi)$ given by the formula 8.957.1 in [3] is used to express $I_m(x,z)$ given by Eq. (9) as

$$I_m(x,z) = \pi^{1/2} w_0 i^m \operatorname*{Lim}_{t=0} \frac{\partial^m}{\partial t^m} \int_{-\infty}^{\infty} dp_x \exp\left(-i2\pi p_x x\right)$$
$$\times \exp\left[-\pi^2 w_0^2 \frac{p_x^2}{q_\pm^2}\right] \exp\left(-t^2 + 2t\sqrt{2}\pi w_0 p_x\right) \tag{AA1}$$

The p_x integration is carried out to yield that

$$\pi^{1/2} w_0 i^m \int_{-\infty}^{\infty} dp_x \exp\left[i2\pi p_x(-x - i\sqrt{2}tw_0)\right] \exp\left[-\pi^2 w_0^2 \frac{p_x^2}{q_\pm^2}\right]$$
$$= q_\pm i^m \exp\left[-\frac{q_\pm^2(x + i\sqrt{2}tw_0)^2}{w_0^2}\right] \tag{AA2}$$

Equation (AA2) is valid only for the two physical spaces $0 < z < \infty$ and $-\infty < z < 0$, where $1/q_\pm^2 \neq 0$. When Eq. (AA2) is substituted into Eq. (AA1), $I_m(x,z)$ simplifies as

$$I_m(x,z) = q_\pm i^m \exp\left(-\frac{q_\pm^2 x^2}{w_0^2}\right) \operatorname*{Lim}_{t=0} \frac{\partial^m}{\partial t^m} \exp\left[t^2 \alpha^2 - 2ti\sqrt{2}x \frac{q_\pm^2}{w_0}\right] \tag{AA3}$$

where

$$\alpha^2 = -1 + 2q_\pm^2 \tag{AA4}$$

Let

$$s = -i\alpha t \tag{AA5}$$

Then, $I_m(x,z)$ is transformed as

$$I_m(x,z) = q_\pm \alpha^m \exp\left(-\frac{q_\pm^2 x^2}{w_0^2}\right) \lim_{s=0} \frac{\partial^m}{\partial s^m} \exp\left[-s^2 + 2s\sqrt{2}\frac{q_\pm^2 x}{\alpha w_0}\right] \tag{AA6}$$

The generating function for $H_m(\xi)$ is used to express the limit as a Hermite function. Equation (AA6) is then determined as

$$I_m(x,z) = q_\pm \alpha^m \exp\left(-\frac{q_\pm^2 x^2}{w_0^2}\right) H_m\left(\frac{\sqrt{2}q_\pm^2 x}{\alpha w_0}\right) \tag{AA7}$$

Substituting $q_\pm$ from Eq. (7) in Eq. (AA4) leads to

$$\alpha = \frac{q_\pm}{q_\pm^*} = \frac{q_\pm}{q_\mp} \tag{AA8}$$

The use of Eq. (AA8) in Eq. (AA7) yields

$$I_m(x,z) = \frac{q_\pm^{m+1}}{q_\pm^{*m}} H_m\left(\frac{\sqrt{2}q_\pm q_\pm^* x}{w_0}\right) \exp\left(-\frac{q_\pm^2 x^2}{w_0^2}\right) \tag{AA9}$$

Appendix AB: Complex space source

One of the factors of the complex space source given by Eq. (20) is found as

$$C_{s,m}(x) = i^m \int_{-\infty}^{\infty} dp_x \exp\left(-i2\pi p_x x\right) H_m(\sqrt{2}\pi w_0 p_x) \tag{AB1}$$

From Eq. (11.3), the Hermite function is replaced by its polynomial representation with the result that

$$C_{s,m}(x) = \sum_{\ell=0}^{\ell=\ell_m} \frac{i^m m!(-1)^\ell 2^{m-2\ell}}{\ell!(m-2\ell)!}(\sqrt{2}\pi w_0)^{m-2\ell} \int_{-\infty}^{\infty} dp_x \exp\left(-i2\pi p_x x\right) p_x^{m-2\ell} \tag{AB2}$$

The term $p_x^{m-2\ell}$ appearing inside the integral can be replaced by the operator $[(i/2\pi)\partial/\partial x]^{m-2\ell}$ placed outside the integral. The remaining integral is $\delta(x)$. Then, $C_{s,m}(x)$ given by Eq. (AB2) is expressed as

$$C_{s,m}(x) = \sum_{\ell=0}^{\ell=\ell_m} a_\ell^m w_0^{m-2\ell} \frac{\partial^{m-2\ell}}{\partial x^{m-2\ell}} \delta(x) \tag{AB3}$$

where

$$a_\ell^m = \frac{m! 2^{(m-2\ell)/2}}{\ell! (m-2\ell)! (-1)^m} \quad \text{for } 0 \le \ell \le \ell_m \tag{AB4}$$

and ℓ_m is the largest integer $\le m/2$. In a similar manner, the second factor of the complex space source as given by Eq. (20) is obtained as

$$C_{s,n}(y) = i^n \int_{-\infty}^{\infty} dp_y \exp\left(-i2\pi p_y y\right) H_n\left(\sqrt{2}\pi w_0 p_y\right) \tag{AB5}$$

Following the same procedure as before, Eq. (AB5) is transformed as

$$C_{s,n}(y) = \sum_{p=0}^{p=p_n} a_p^n w_0^{n-2p} \frac{\partial^{n-2p}}{\partial y^{n-2p}} \delta(y) \tag{AB6}$$

where

$$a_p^n = \frac{n! 2^{(n-2p)/2}}{p! (n-2p)! (-1)^n} \quad \text{for } 0 \le p \le p_n \tag{AB7}$$

and p_n is the largest integer $\le n/2$. Using Eqs. (AB3) and (AB6), the complex space source given by Eq. (20) can be expressed as

$$C_{s,mn}(x,y,z) = \frac{N_{mn}}{ik} \pi w_0^2 \sum_{\ell=0}^{\ell=\ell_m} a_\ell^m w_0^{m-2\ell} \frac{\partial^{m-2\ell}}{\partial x^{m-2\ell}} \delta(x) \sum_{p=0}^{p=p_n} a_p^n w_0^{n-2p} \frac{\partial^{n-2p}}{\partial y^{n-2p}} \delta(y) \tag{AB8}$$

Thus, the complex space source located at $|z| - ib = 0$ consists of a finite series of higher-order point sources.

References

1. H. Kogelnik and T. Li, "Laser beams and resonators," *Appl. Opt.* **5**, 1550–1567 (1966).

2. D. Marcuse, *Light Transmission Optics* (Van Nostrand Reinhold, New York, 1972), Chap. 6.

3. I. S. Gradshteyn and I. M. Ryzhik, *Table of Integrals, Series and Products* (Academic Press, New York, 1965).

CHAPTER 13

Basic full modified Bessel–Gauss wave

For a modified Bessel–Gauss beam, the modulating function for the Gaussian is a Bessel function of imaginary argument or a modified Bessel function of real argument. The scalar modified Bessel–Gauss beam was introduced and its basic full-wave generalization was treated with particular emphasis on its spreading properties on propagation [1]. One form of the electromagnetic modified Bessel–Gauss beams, namely the transverse magnetic (TM) modified Bessel–Gauss beam, was investigated and its basic full-wave generalization was obtained [2]. To generate the TM modified Bessel–Gauss beam and wave, a single component of the magnetic vector potential in the direction of propagation was used. In this chapter, a linearly polarized electromagnetic modified Bessel–Gauss beam and a wave are treated. To produce the linearly polarized modified Bessel–Gauss beam and a wave, a single component of the magnetic vector potential perpendicular to the direction of propagation is required. The complex power associated with the linearly polarized modified Bessel–Gauss beam is deduced. The reactive power of the paraxial beam vanishes. The source in the complex space required for the basic full-wave generalization of the electromagnetic modified Bessel–Gauss beam is obtained. The basic full modified Bessel–Gauss wave generated by the source in the complex space is determined [3,4]. The real and reactive powers of the basic full modified Bessel–Gauss wave are evaluated. As for the previously treated basic full Gaussian waves, the reactive power of the modified Bessel–Gauss wave is infinite. The general characteristics of the real power are discussed as functions of kw_0 and αw_0, where α is the beam shape parameter. The real power increases, approaching the limiting value of the paraxial beam for all values of αw_0 as kw_0 is increased.

1 Modified Bessel–Gauss beam

1.1 Paraxial beam

A linearly polarized electromagnetic modified Bessel–Gauss beam is generated by the x component of the magnetic vector potential at the secondary source plane $z = 0$, as given by

$$a_{x0}^{\pm}(\rho, \phi, 0) = A_{x0}^{\pm}(\rho, \phi, 0) = \frac{N_{\text{MB}}}{ik} \cos m\phi \exp(-\nu)$$

$$\times I_m(\alpha\rho) \exp(-\rho^2/w_0^2) \tag{1}$$

where k is the wavenumber, N_{MB} is a normalization constant, m is the azimuthal mode number, w_0 is the beam waist at the focal plane $z = 0$, $\nu = (\alpha w_0/2)^2$, α is the beam shape parameter, and $I_m(\ )$ is the modified Bessel function of order m. It is possible for the normalization constant to depend on the mode number m. The medium in $|z| > 0$ is cylindrically symmetrical about the z axis. Therefore, the ϕ dependence of $a_{x0}^{\pm}(\rho, \phi, z)$ and $A_{x0}^{\pm}(\rho, \phi, z)$ is the same $\cos m\phi$ as at the input plane $z = 0$. The mode number is a positive integer starting with 0.

The paraxial approximation of the x component of the magnetic vector potential is given by $A_{x0}^{\pm}(\rho, \phi, z)$. The slowly varying amplitude $a_{x0}^{\pm}(\rho, \phi, z)$ is related to $A_{x0}^{\pm}(\rho, \phi, z)$ as stated in Eq. (9.5). The paraxial wave equation satisfied by $a_{x0}^{\pm}(\rho, \phi, z)$ and the differential equation satisfied by the mth order Bessel transform $\overline{a}_{x0}^{\pm}(\eta, \phi, z)$ of $a_{x0}^{\pm}(\rho, \phi, z)$ are as expressed in Eqs. (9.6) and (9.7), respectively. The solution of $\overline{a}_{x0}^{\pm}(\eta, \phi, z)$ in terms of $\overline{a}_{x0}^{\pm}(\eta, \phi, 0)$, as obtained in Eq. (9.8), is

$$\overline{a}_{x0}^{\pm}(\eta, \phi, z) = \overline{a}_{x0}^{\pm}(\eta, \phi, 0) \exp\left(-\frac{\eta^2 w_0^2}{4} \frac{i|z|}{b}\right) \tag{2}$$

From Eq. (1), $\overline{a}_{x0}^{\pm}(\eta, \phi, 0)$ and $\overline{A}_{x0}^{\pm}(\eta, \phi, 0)$ are obtained as

$$\overline{a}_{x0}^{\pm}(\eta, \phi, 0) = \overline{A}_{x0}^{\pm}(\eta, \phi, 0) = \frac{N_{\text{MB}}}{ik} \cos m\phi \exp(-\nu)$$

$$\times \int_0^\infty d\rho\rho J_m(\eta\rho) I_m(\alpha\rho) \exp(-\rho^2/w_0^2) \tag{3}$$

Using formula 6.633.4 in [5], the integral is evaluated, with the result that

$$\overline{a}_{x0}^{\pm}(\eta, \phi, 0) = \frac{N_{\text{MB}}}{ik} \cos m\phi \frac{w_0^2}{2}$$

$$\times J_m(\eta d_\alpha) \exp(-\eta^2 w_0^2/4) \tag{4}$$

where $J_m(\ )$ is the Bessel function of order m and

$$d_\alpha = \frac{1}{2}\alpha w_0^2 \tag{5}$$

When Eq. (4) is substituted into Eq. (2), $\overline{a}_{x0}^{\pm}(\eta, \phi, z)$ is obtained as

$$\overline{a}_{x0}^{\pm}(\eta, \phi, z) = \frac{N_{\text{MB}}}{ik} \cos m\phi \frac{w_0^2}{2} J_m(\eta d_\alpha) \exp\left(-\frac{\eta^2 w_0^2}{4q_{\pm}^2}\right) \tag{6}$$

where

$$q_{\pm} = \left(1 \pm \frac{iz}{b}\right)^{-1/2} \tag{7}$$

and $b = \frac{1}{2}kw_0^2$. The sign $\pm$, the superscript $\pm$, and the subscript $\pm$ indicate that the propagation is in the $\pm z$ direction.

The inverse Bessel transform of Eq. (6) is given by

$$a_{x0}^{\pm}(\rho,\phi,z) = \frac{N_{MB}}{ik}\cos m\phi \frac{w_0^2}{2}\int_0^{\infty} d\eta\eta \, J_m(\eta\rho)J_m(\eta d_a)\exp\left(-\frac{\eta^2 w_0^2}{4q_{\pm}^2}\right) \tag{8}$$

Using formula 6.633.2 in [5], the integral in Eq. (8) is evaluated to yield

$$a_{x0}^{\pm}(\rho,\phi,z) = \frac{N_{MB}}{ik}\cos m\phi \; q_{\pm}^2\exp\left(-q_{\pm}^2 v\right) I_m(q_{\pm}^2 a\rho)\exp\left(-\frac{q_{\pm}^2\rho^2}{w_0^2}\right) \tag{9}$$

The rapidly varying phase factor $\exp(\pm ikz)$ is included, and from Eq. (9) the vector potential governing the linearly polarized electromagnetic modified Bessel–Gauss beam is obtained as

$$A_{x0}^{\pm}(\rho,\phi,z) = \exp(\pm ikz)\frac{N_{MB}}{ik}\cos m\phi \; q_{\pm}^2\exp\left(-q_{\pm}^2 v\right)$$

$$\times I_m(q_{\pm}^2 a\rho)\exp\left(-\frac{q_{\pm}^2\rho^2}{w_0^2}\right) \tag{10}$$

$A_{x0}^{\pm}(\rho,\phi,z)$, given by Eq. (10), is valid only for the position coordinates in the two physical spaces $0 < z < \infty$ and $-\infty < z < 0$, where $1/q_{\pm}^2 \neq 0$. From Eq. (9.5), it is seen that

$$\overline{A}_{x0}^{\pm}(\eta,\phi,z) = \exp(\pm ikz)\overline{a}_{x0}^{\pm}(\eta,\phi,z) \tag{11}$$

The modified Bessel–Gauss beam has a closed-form Bessel transform representation, as given by Eqs. (6) and (11).

1.2 Time-averaged power

The electromagnetic fields, their characteristics, the induced electric current density on the secondary source plane $z = 0$, and the complex power associated with the modified Bessel–Gauss beam are the same as those for the complex-argument Laguerre–Gauss beam. $A_{x0}^{\pm}(\rho,\phi,z)$ and $a_{x0}^{\pm}(\rho,\phi,z)$ corresponding to the modified Bessel–Gauss beam have to be used. As given in Section 9.1, the reactive power of the modified Bessel–Gauss beam vanishes; that is,

$$P_{im} = 0 \tag{12}$$

Substituting $\overline{a}_{x0}^{\pm}(\eta,\phi,0)$ from Eq. (4) into Eq. (9.23) and carrying out the ϕ integration, the real power is obtained as

$$P_{re} = c\pi\varepsilon_m \frac{N_{MB}^2 w_0^4}{4}\int_0^{\infty} d\eta\eta J_m^2(\eta d_a)\exp\left(-\frac{\eta^2 w_0^2}{2}\right) \tag{13}$$

The integral in Eq. (13) is evaluated using formula 6.633.2 in [5], with the result that

$$P_{re} = c\pi\varepsilon_m \frac{N_{MB}^2 w_0^2}{4} \exp(-\nu) I_m(\nu) \tag{14}$$

The normalization constant is chosen as

$$N_{MB} = \left[\frac{8}{c\pi w_0^2 \varepsilon_m \exp(-\nu) I_m(\nu)} \right]^{1/2} \tag{15}$$

From Eqs. (14) and (15), P_{re} is found as

$$P_{re} = 2 \text{ W} \tag{16}$$

The time-averaged power transported by the paraxial beam in the $\pm z$ direction is denoted by $P_{MB}^{\pm}$. By symmetry, $P_{MB}^+ = P_{MB}^-$. The real power is the time-averaged power created by the current source. One half of this power is transported in the $+z$ direction and the other half in the $-z$ direction. From Eq. (16), it is obtained that

$$P_{MB}^+ = P_{MB}^- = \frac{1}{2} P_{re} = 1 \text{ W} \tag{17}$$

The normalization constant N_{MB} is chosen such that the time-averaged power $P_{MB}^{\pm}$ carried by the paraxial beam in the $\pm z$ direction is given by $P_{MB}^{\pm} = 1$ W.

In the limit of $\alpha = \nu = 0$, the modified Bessel–Gauss beam reduces to the corresponding Laguerre–Gauss beam of radial mode number $n = 0$. For $n = 0$, the real-argument and complex-argument Laguerre–Gauss beams are identical. From Eq. (9.26), for $n = 0$, the normalization constant of the Laguerre–Gauss beam is obtained as

$$N_{0m} = \left(\frac{8}{c\pi w_0^2 \varepsilon_m m!} \right)^{1/2} \tag{18}$$

By use of the small argument approximation of $I_m(\nu)$ from formula 8.445 in [5] and the relation $\nu = (\alpha w_0/2)^2$, the normalization constant N_{MB} of the modified Bessel–Gauss beam, in the limit of $\alpha = \nu = 0$, is expressed in terms of N_{0m} given by Eq. (18) as

$$N_{MB} = N_{0m} \frac{2^{m/2} 2^m m!}{\alpha^m w_0^m} \tag{19}$$

As anticipated, Eq. (19) reveals that the normalization constant N_{MB} is a function of the mode number m. In Eq. (10), $I_m(q_{\pm}^2 \alpha \rho)$ is replaced by its small argument approximation and N_{MB} is replaced by its value given in Eq. (19). Then, in the limit of $\alpha = \nu = 0$, $A_{x0}^{\pm}(\rho, \phi, z)$ of the modified Bessel–Gauss beam is determined from Eq. (10) as

$$A_{x0}^{\pm}(\rho, \phi, z) = \exp(\pm ikz) \frac{N_{0m}}{ik} 2^{m/2} q_{\pm}^{m+2} \cos m\phi$$

$$\times \left(\frac{q_{\pm}^2 \rho^2}{w_0^2} \right)^{m/2} \exp\left(-\frac{q_{\pm}^2 \rho^2}{w_0^2} \right) \quad \text{for } \alpha = 0 \tag{20}$$

For $n = 0, A_{x0}^{\pm}(\rho, \phi, z)$ of the Laguerre–Gauss beam, as given by Eq. (9.16), becomes identical to that given by Eq. (20). Thus, in the limit of $\alpha = \nu = 0$, the modified Bessel–Gauss beam becomes identical to the corresponding Laguerre–Gauss beam of radial mode number $n = 0$.

2 Modified Bessel–Gauss wave

For the basic full-wave generalization of the modified Bessel–Gauss beam, the search for the complex space source is made at $|z| - ib = 0$. Then, $1/q_{\pm}^2 = 0$, and from Eq. (8) the complex space source is obtained as

$$C_{s,m}(\rho, \phi, z) = \frac{N_{\mathrm{MB}}}{ik} \cos m\phi \frac{w_0^2}{2} \int_0^\infty d\eta\eta J_m(\eta\rho) J_m(\eta d_\alpha) \tag{21}$$

$$= \frac{N_{\mathrm{MB}}}{ik} \cos m\phi \frac{w_0^2}{2} \frac{\delta(\rho - d_\alpha)}{\rho} \tag{22}$$

Equation (9.40) is used in finding the result given by Eq. (22). The complex space source in the form given by Eq. (21) is adequate for deriving the expression for the full wave. The source given by Eq. (21) located at $|z| - ib = 0$ gives rise to the paraxial beam for $|z| > 0$. The same source situated at $|z| = 0$ generates only the asymptotic limit of the paraxial beam. The paraxial equation is only an approximation for the Helmholtz equation. Therefore, the complex space source derived from the paraxial beam, except for an excitation coefficient S_{ex}, is used for the Helmholtz equation to deduce the asymptotic limit of the full wave. Hence, the source that produces the asymptotic limit of the full-wave solution of the Helmholtz equation is determined from Eq. (21) as

$$C_{s\infty,m}(\rho, \phi, z) = S_{ex} \frac{N_{\mathrm{MB}}}{ik} \cos m\phi \frac{w_0^2}{2} \delta(z) \int_0^\infty d\eta\eta J_m(\eta\rho) J_m(\eta d_\alpha) \tag{23}$$

For the source given by Eq. (23), let $G_m(\rho, \phi, z)$ be the solution of the reduced Helmholtz equation. Then, $G_m(\rho, \phi, z)$ is governed by the following differential equation:

$$\left(\frac{\partial^2}{\partial\rho^2} + \frac{1}{\rho}\frac{\partial}{\partial\rho} - \frac{m^2}{\rho^2} + \frac{\partial^2}{\partial z^2} + k^2 \right) G_m(\rho, \phi, z) = -S_{ex} \frac{N_{\mathrm{MB}}}{ik} \cos m\phi \frac{w_0^2}{2} \delta(z)$$
$$\times \int_0^\infty d\eta\eta J_m(\eta\rho) J_m(\eta d_\alpha) \tag{24}$$

The differential equation satisfied by the mth order Bessel transform $\overline{G}_m(\eta, \phi, z)$ of $G_m(\rho, \phi, z)$ is found from Eq. (24) as

$$\left(\frac{\partial^2}{\partial z^2} + \zeta^2 \right) \overline{G}_m(\eta, \phi, z) = -S_{ex} \frac{N_{\mathrm{MB}}}{ik} \cos m\phi \frac{w_0^2}{2} J_m(\eta d_\alpha)\delta(z) \tag{25}$$

where ζ is the same as that given by Eq. (9.45). The solution of Eq. (25) is given by

$$\overline{G}_m(\eta, \phi, z) = \frac{iS_{ex}}{2} \frac{N_{MB}}{ik} \cos m\phi \frac{w_0^2}{2} J_m(\eta d_a) \, \zeta^{-1} \exp\left(i\zeta|z|\right) \tag{26}$$

$G_m(\rho, \phi, z)$ is obtained from Eq. (26) by the inverse Bessel transformation. $A_x^\pm(\rho, \phi, z)$ is deduced from $G_m(\rho, \phi, z)$ by the analytical continuation from $|z|$ to $|z| - ib$. The paraxial approximation $A_{x0}^\pm(\rho, \phi, z)$ of $A_x^\pm(\rho, \phi, z)$ is then found. When the paraxial approximation $A_{x0}^\pm(\rho, \phi, z)$ is compared with Eq. (8) after the inclusion of the phase factor $\exp(\pm ikz)$, $A_x^\pm(\rho, \phi, z)$ is observed to reduce correctly to the initially chosen paraxial beam if S_{ex} is chosen to be the same as that given by Eq. (9.48). This value of S_{ex} is inserted into the expression for $A_x^\pm(\rho, \phi, z)$. The final expression for $A_x^\pm(\rho, \phi, z)$ is given by

$$A_x^\pm(\rho, \phi, z) = k \exp(-kb) \frac{N_{MB}}{ik} \cos m\phi \frac{w_0^2}{2} \int_0^\infty d\eta\eta J_m(\eta\rho) J_m(\eta d_a) \zeta^{-1} \exp\left[i\zeta(|z| - ib)\right] \tag{27}$$

The full wave represented by Eq. (27) is designated the basic full modified Bessel–Gauss wave.

3 Real and reactive powers

A comparison of Eq. (9.49) with Eq. (27) reveals that by changing $N_{nm}(-1)^n 2^{-m/2}(\eta w_0)^{2n+m}$ to $N_{MB}J_m(\eta d_a)$, the vector potential, the electromagnetic fields, and the source current density of the complex-argument Laguerre–Gauss wave are transformed to the corresponding quantities pertaining to the modified Bessel–Gauss wave. Therefore, from Eq. (9.64), the expression for the complex power of the modified Bessel–Gauss wave is obtained as

$$\begin{aligned}
P_C = -\frac{c}{2} \int_0^\infty d\rho\rho \int_0^{2\pi} d\phi k \, \exp(-kb) N_{MB} \frac{w_0^2}{2} \cos m\phi \int_0^\infty d\eta\eta J_m(\eta\rho) \\
\times \left(1 - \frac{\gamma_m \eta^2}{2k^2}\right) J_m(\eta d_a) \zeta^{-1} \exp(\zeta b)(-2) \exp(-kb) N_{MB} \frac{w_0^2}{2} \cos m\phi \\
\times \int_0^\infty d\overline{\eta}\overline{\eta} J_m(\overline{\eta}\rho) J_m(\overline{\eta} d_\varepsilon) \exp(\overline{\zeta}^* b)
\end{aligned} \tag{28}$$

The ϕ integration is performed, resulting in $\pi\varepsilon_m$, where $\varepsilon_m = 1$ for $m \neq 0$ and $\varepsilon_m = 2$ for $m = 0$. The ρ integration is carried out using Eq. (9.22) to yield $\delta(\eta - \overline{\eta})/\eta$. Then, the $\overline{\eta}$ integration is performed. The result is

$$P_C = ck \exp(-2kb) \frac{N_{MB}^2 w_0^4}{4} \pi\varepsilon_m \int_0^\infty d\eta\eta \left(1 - \frac{\gamma_m \eta^2}{2k^2}\right) J_m^2(\eta d_a) \zeta^{-1} \exp\left[b(\zeta + \zeta^*)\right] \tag{29}$$

where γ_m is defined by Eq. (9.62).

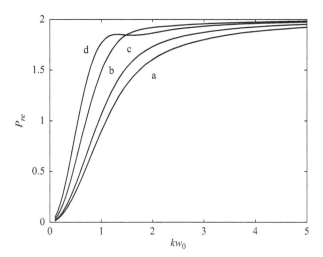

Fig. 1.   Real power P_{re} as a function of kw_0 for $0.1 < kw_0 < 5$, the azimuthal mode number $m = 0$, and (a) $aw_0 = 0$, (b) $aw_0 = 1$, (c) $aw_0 = 2$, and (d) $aw_0 = 3$. The normalization is such that the real power in the corresponding paraxial beam is 2 W.

The reactive power P_{im}, the imaginary part of P_C, arises from η in the range $k < \eta < \infty$, where ζ is imaginary. Therefore, from Eqs. (29) and (9.45), P_{im} is determined as

$$P_{im} = -ck \exp(-2kb) \frac{N_{\mathrm{MB}}^2 w_0^4}{4} \pi \varepsilon_m \int_k^\infty d\eta\eta \left(1 - \frac{\gamma_m \eta^2}{2k^2} \right) J_m^2(\eta d_a)(\eta^2 - k^2)^{-1/2} \qquad (30)$$

From the asymptotic expansion of $J_m(\eta d_a)$, as given by formula 3.5c in [6], for very large η, the integral in Eq. (30) is approximated as

$$I_\infty = -\frac{\gamma_m}{\pi k^2 d_a} \int_k^\infty d\eta\eta\cos^2\left(\eta d_a - \frac{m\pi}{2} - \frac{\pi}{4} \right) = \infty \qquad (31)$$

The value of the integral is infinite, as indicated in Eq. (31). Therefore,

$$P_{im} = \infty \qquad (32)$$

The reactive power P_{im} associated with the basic full modified Bessel–Gauss wave is infinite. This result is to be expected since the complex space source [Eq. (22)] generating the full wave is a filamentary circular current loop of radius d_α and vanishingly small thickness. In order to obtain finite values of the reactive power, it is necessary to have a complex space source of finite dimensions.

The real power P_{re}, the real part of P_C, arises from η in the range $0 < \eta < k$, where ζ is real. From Eqs. (29) and (9.45), P_{re} is obtained as

$$P_{re} = \frac{2kw_0^2}{\exp(-\nu)I_m(\nu)} \int_0^k d\eta\eta \left(1 - \frac{\gamma_m \eta^2}{2k^2} \right) J_m^2(\eta d_a)(k^2 - \eta^2)^{-1/2}\exp\left[-2b(k - \zeta)\right] \qquad (33)$$

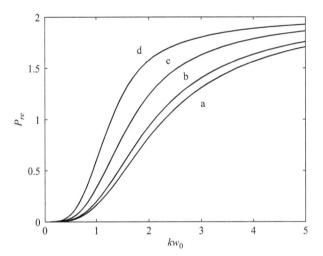

Fig. 2.   Same as in Fig. 1, but for the azimuthal mode number $m = 1$.

The value of N_{MB} is substituted from Eq. (15) in obtaining Eq. (33). Therefore, the normalization of the wave amplitude is such that the real power in the corresponding paraxial beam is $P_{re} = 2$ W, as given by Eq. (16). The integration variable is modified as $\eta = k\sin\theta$. Then, Eq. (33) is transformed as

$$P_{re} = \frac{2k^2 w_0^2}{\exp(-v)I_m(v)} \int_0^{\pi/2} d\theta \, \sin\theta \left(1 - \frac{\gamma_m}{2}\sin^2\theta\right)$$
$$\times J_m^2\left(\frac{1}{2}kw_0\alpha w_0\sin\theta\right)\exp\left[-k^2 w_0^2(1-\cos\theta)\right] \tag{34}$$

In the limit of $\alpha w_0/2 = v^{1/2} = 0$, Eq. (34) reduces to Eq. (9.70) for $n = 0$. Thus, the real power of the basic full modified Bessel–Gauss wave is identical to the real power of the Laguerre–Gauss wave with the radial mode number $n = 0$.

The real power P_{re} depends on kw_0, m, and αw_0. The real power P_{re} is shown in Fig. 1 as a function of kw_0 for $0.1 < kw_0 < 5$, for $\alpha w_0 = 0$, 1, 2, and 3, and for the azimuthal mode number $m = 0$. In Fig. 2, P_{re} is depicted as a function of kw_0 for $m = 1$; the other parameters are the same as in Fig. 1. In general, the real power increases, approaching the limiting value of the paraxial beam as kw_0 is increased.

References

1. S. R. Seshadri, "Scalar modified Bessel-Gauss beams and waves," *J. Opt. Soc. Am. A* **24**, 2837–2842 (2007).

2. S. R. Seshadri, "Electromagnetic modified Bessel-Gauss beams and waves," *J. Opt. Soc. Am. A* **25**, 1–8 (2008).

3. S. R. Seshadri, "Virtual source for the Bessel-Gauss beam," *Opt. Lett.* **27**, 998–1000 (2002).

4. S. R. Seshadri, "Quality of paraxial electromagnetic beams," *Appl. Opt.* **45**, 5335–5345 (2006).

5. I. S. Gradshteyn and I. M. Ryzhik, *Table of Integrals, Series and Products* (Academic Press, New York, 1965).

6. W. Magnus and F. Oberhettinger, *Functions of Mathematical Physics* (Chelsea Publishing Company, New York, 1954).

CHAPTER 14

Partially coherent and partially incoherent full Gaussian wave

The fundamental Gaussian beam treated in Chapter 1 is an approximate solution of the governing equation [1]. A class of full-wave generalizations of the fundamental Gaussian beam known as the extended full Gaussian wave has been obtained [2–4] and is treated in Chapter 6. A virtual source in the complex space is deduced for developing the extended full Gaussian wave. The actual secondary source in the physical space, which is entirely equivalent to the virtual source in the complex space, is derived for determining completely the dynamics of the extended full Gaussian wave [4]. The secondary source is an infinitesimally thin planar sheet of current that launches the outwardly propagating waves in the two half spaces separating the current sheet. The sources, the potentials, and the fields have a harmonic time dependence with wave period T_w.

Wolf et al. [5–7] introduced the idea of planar secondary sources for the treatment of partially coherent light beams. For the extended full Gaussian waves, only the fully coherent waves in which the wave amplitude remains a constant in time were considered [3]. For the partially coherent beams, the wave period is the same, namely T_w, but the source amplitude is essentially a constant on the time scale of T_w; but on a longer time scale T_f, on the order of nearly thousands of T_w, the amplitude changes in a random manner [8]. A majority of treatments of the partially coherent beams are restricted to the paraxial beams for which the beam waist is large compared to the wavelength. There is a need for the treatment of partially coherent, spatially localized electromagnetic waves extended beyond the paraxial approximation to the full waves governed by Maxwell's equations. An analysis of the partially coherent, spatially localized electromagnetic waves was presented for the fundamental Gaussian wave [9,10]. This is a special case ($b_t/b = 0$) for which the virtual source becomes identical to the actual secondary source in the physical space. In this chapter, the treatment of partially coherent, spatially localized electromagnetic waves is enlarged in scope to include the extended full Gaussian waves for which the virtual source is in the complex space requiring a different formulation.

The complex space source theory is used to derive an integral expression for the vector potential that generates the extended full Gaussian wave in terms of the input value of the vector potential of the corresponding paraxial beam. The vector potential and the fields are assumed to fluctuate on a time scale on the order of T_f, which is very large compared to T_w.

161

The Poynting vector in the propagation direction averaged over a wave period is expressed in terms of the cross-spectral density of the fluctuating vector potential across the input plane. The partially coherent, spatially localized electromagnetic wave is completely specified by the input value of the cross-spectral density, that is, the correlations of the electromagnetic fields between two points in the source plane [7–10]. The original Schell model source [11,12] is assumed for the cross-spectral density for the partially coherent waves, that is, fully coherent and nearly fully coherent waves. For the partially incoherent waves (that is, fully incoherent and nearly fully incoherent waves), a modified Schell model source is used for the cross-spectral density. The modified Schell model source reproduces as a limiting case the delta function cross-spectral density introduced by Goodman [13] for the fully incoherent source. The radiation intensity and the radiated power are determined. The effect of spatial coherence on the radiation intensity distribution and the radiated power are investigated [14].

1 Extended full-wave generalization of the paraxial beam

An infinitesimally thin sheet of electric current is located on the secondary source plane $z = 0$. The wave generated by the source propagates in the $+z$ direction in $z > 0$ and in the $-z$ direction in $z < 0$. The source and the fields have a harmonic time dependence of the form $\exp(-i\omega t)$, where $\omega/2\pi$ is the wave frequency. The source electric current density is in the x direction. The generated magnetic vector potential $A_x^{\pm}(x,y,z)$ is also in the x direction. The paraxial approximation of $A_x^{\pm}(x,y,z)$ is given by $A_{x0}^{\pm}(x,y,z)$, where the subscript 0 stands for paraxial. The slowly varying amplitude $a_{x0}^{\pm}(x,y,z)$ of $A_{x0}^{\pm}(x,y,z)$ is defined by

$$A_{x0}^{\pm}(x,y,z) = \exp(\pm ikz)\, a_{x0}^{\pm}(x,y,z) \tag{1}$$

where $\exp(\pm ikz)$ is the rapidly varying phase factor and k is the wavenumber. The sign $\pm$ and the superscript $\pm$ denote that the propagation is in the $\pm z$ direction. The differential equation satisfied by $a_{x0}^{\pm}(x,y,z)$ is given by

$$\left(\frac{\partial^2}{\partial x^2} + \frac{\partial^2}{\partial y^2} \pm 2ik\frac{\partial}{\partial z}\right) a_{x0}^{\pm}(x,y,z) = 0 \tag{2}$$

$A_x^{\pm}(x,y,z)$, $A_{x0}^{\pm}(x,y,z)$, and $a_{x0}^{\pm}(x,y,z)$ are all assumed to have a Gaussian-type distribution as a factor. The e-folding distance of the Gaussian part of the distribution is the beam waist w_0. The Rayleigh distance is $b = \frac{1}{2}kw_0^2$. The Fourier transform representation $\bar{a}_{x0}^{\pm}(p_{x1},p_{y1},z)$ [Eq. (A17)] of $a_{x0}^{\pm}(x,y,z)$ is substituted into Eq. (2) to find the differential equation satisfied by $\bar{a}_{x0}^{\pm}(p_{x1},p_{y1},z)$. The solution of this differential equation is

$$\bar{a}_{x0}^{\pm}(p_{x1},p_{y1},z) = \bar{a}_{x0}^{\pm}(p_{x1},p_{y1},z=0)\exp\left[-\pi^2 w_0^2\left(p_{x1}^2 + p_{y1}^2\right)\frac{i|z|}{b}\right] \tag{3}$$

The input value of the paraxial approximation of the vector potential is $a_{x0}^{\pm}(x_1,y_1,0)$, which is the same as $A_{x0}^{\pm}(x_1,y_1,0)$. The Fourier transform $\bar{a}_{x0}^{\pm}(p_{x1},p_{y1},0)$ [Eq. (A18)] of $a_{x0}^{\pm}(x_1,y_1,0)$ is substituted into Eq. (3), and the resulting $\bar{a}_{x0}^{\pm}(p_{x1},p_{y1},z)$ is inverted to obtain

$a_{x0}^{\pm}(x, y, z)$. Inserting the rapidly varying plane-wave phase factor, as given in Eq. (1), the paraxial approximation of the magnetic vector potential is found as

$$A_{x0}^{\pm}(x, y, z) = \exp(\pm ikz) \int_{-\infty}^{\infty} \int_{-\infty}^{\infty} \int_{-\infty}^{\infty} \int_{-\infty}^{\infty} dx_1 dy_1 dp_{x1} dp_{y1}$$

$$\times a_{x0}^{\pm}(x_1, y_1, z_1 = 0) \exp[i2\pi(p_{x1}x_1 + p_{y1}y_1)]$$

$$\times \exp\left[-\pi^2 w_0^2 \left(p_{x1}^2 + p_{y1}^2\right) \frac{i|z|}{b}\right] \exp[-i2\pi(p_{x1}x + p_{y1}y)] \qquad (4)$$

The search for the complex space source is made at $|z| = ib_t$, where the length parameter b_t lies in the range $0 \leq b_t \leq b$. Therefore, the source for the paraxial beam is $a_{x0}^{\pm}(x, y, |z| = ib_t)$. The asymptotic value of the paraxial beam corresponding to $|z| \to \infty$ is obtained by moving the source, $a_{x0}^{\pm}(x, y, |z| = ib_t)$, to $|z| = 0$. Since the paraxial wave equation is only an approximation for the exact Helmholtz equation, the same source $a_{x0}^{\pm}(x, y, |z| = ib_t)$, with the inclusion of the excitation coefficient S_{ex} as a factor, placed at $|z| = 0$ and used for the Helmholtz equation should yield the asymptotic value of the exact vector potential $A_x^{\pm}(x, y, z)$. Let $G(x, y, z)$ represent the asymptotic value of the exact vector potential. Then, $G(x, y, z)$ satisfies the following differential equation:

$$\left(\frac{\partial^2}{\partial x^2} + \frac{\partial^2}{\partial y^2} + \frac{\partial^2}{\partial z^2} + k^2\right) G(x, y, z) = -S_{ex}\delta(z) a_{x0}^{\pm}(x, y, |z| = ib_t) \qquad (5)$$

Substituting the Fourier integral representation $\overline{G}(p_{x1}, p_{y1}, z)$ of $G(x, y, z)$ into Eq. (5) leads to the differential equation satisfied by $\overline{G}(p_{x1}, p_{y1}, z)$ as

$$\left(\frac{\partial^2}{\partial z^2} + \zeta_1^2\right) \overline{G}(p_{x1}, p_{y1}, z) = -S_{ex}\delta(z) \overline{a}_{x0}^{\pm}(p_{x1}, p_{y1}, |z| = ib_t)$$

$$= -S_{ex}\delta(z) \int_{-\infty}^{\infty} \int_{-\infty}^{\infty} dx_1 dy_1 a_{x0}^{\pm}(x_1, y_1, z_1 = 0)$$

$$\times \exp\left[i2\pi(p_{x1}x_1 + p_{y1}y_1)\right] \exp\left[\pi^2 w_0^2 \left(p_{x1}^2 + p_{y1}^2\right) \frac{b_t}{b}\right] \qquad (6)$$

where

$$\zeta_n = \left[k^2 - 4\pi^2 \left(p_{xn}^2 + p_{yn}^2\right)\right]^{1/2} \quad \text{for } n = 1, 2 \qquad (7)$$

The solution of the differential equation given by Eq. (6) is obtained as [see Eqs. (A19) and (A25)]

$$\overline{G}(p_x, p_y, z) = \frac{iS_{ex}}{2} \int_{-\infty}^{\infty} \int_{-\infty}^{\infty} dx_1 dy_1 a_{x0}^{\pm}(x_1, y_1, z_1 = 0)$$

$$\times \exp\left[i2\pi(p_{x1}x_1 + p_{y1}y_1)\right] \exp\left[\pi^2 w_0^2 \left(p_{x1}^2 + p_{y1}^2\right) \frac{b_t}{b}\right]$$

$$\times \zeta_1^{-1} \exp(i\zeta_1 |z|) \qquad (8)$$

From Eq. (8), by inverse Fourier transformation, $G(x, y, z)$ is obtained as

$$
G(x, y, z) = \frac{iS_{ex}}{2} \int_{-\infty}^{\infty} \int_{-\infty}^{\infty} \int_{-\infty}^{\infty} \int_{-\infty}^{\infty} dx_1 dy_1 dp_{x1} dp_{y1}
$$
$$
\times a_{x0}^{\pm}(x_1, y_1, z_1 = 0) \exp\left[i2\pi(p_{x1}x_1 + p_{y1}y_1)\right]
$$
$$
\times \exp\left[\pi^2 w_0^2 \left(p_{x1}^2 + p_{y1}^2\right) \frac{b_t}{b}\right] \exp\left[-i2\pi(p_{x1}x + p_{y1}y)\right]
$$
$$
\times \zeta_1^{-1} \exp(i\zeta_1 |z|) \tag{9}
$$

By analytically continuing $G(x, y, z)$ from $|z|$ to $(|z| - ib_t)$, the exact vector potential is determined as

$$
A_x^{\pm}(x, y, z) = \frac{iS_{ex}}{2} \int_{-\infty}^{\infty} \int_{-\infty}^{\infty} \int_{-\infty}^{\infty} \int_{-\infty}^{\infty} dx_1 dy_1 dp_{x1} dp_{y1}
$$
$$
\times a_{x0}^{\pm}(x_1, y_1, z_1 = 0) \exp[i2\pi(p_{x1}x_1 + p_{y1}y_1)]
$$
$$
\times \exp\left[\pi^2 w_0^2 (p_{x1}^2 + p_{y1}^2) \frac{b_t}{b}\right] \exp[-i2\pi(p_{x1}x + p_{y1}y)]
$$
$$
\times \zeta_1^{-1} \exp[i\zeta_1(|z| - ib_t)] \tag{10}
$$

In the paraxial approximation, $k^2 \gg 4\pi^2(p_{x1}^2 + p_{y1}^2)$; in the amplitude term, ζ_1 is approximated by k; and in the phase term, ζ_1 is approximated by

$$
\zeta_1 = k - \frac{\pi^2 w_0^2}{b}(p_{x1}^2 + p_{y1}^2) \tag{11}
$$

Therefore, from Eq. (10), the paraxial approximation is determined as

$$
A_{x0}^{\pm}(x, y, z) = \exp(\pm ikz)\frac{iS_{ex}}{2k} \exp(kb_t) \int_{-\infty}^{\infty} \int_{-\infty}^{\infty} \int_{-\infty}^{\infty} \int_{-\infty}^{\infty} dx_1 dy_1 dp_{x1} dp_{y1}
$$
$$
\times a_{x0}^{\pm}(x_1, y_1, z_1 = 0) \exp[i2\pi(p_{x1}x_1 + p_{y1}y_1)]
$$
$$
\times \exp\left[-\pi^2 w_0^2(p_{x1}^2 + p_{y1}^2)\frac{i|z|}{b}\right] \exp[-i2\pi(p_{x1}x + p_{y1}y)] \tag{12}
$$

The excitation coefficient is chosen as

$$
S_{ex} = -2ik \exp(-kb_t) \tag{13}
$$

Then, Eq. (12), together with Eq. (13), is in agreement with Eq. (4). Therefore, the exact vector potential given by Eq. (10) is in agreement with Eq. (4) in the paraxial limit if S_{ex} is chosen in accordance with Eq. (13). Substituting Eq. (13) into Eq. (10), the exact vector potential is found as

$$A_x^{\pm}(x,y,z) = k \exp(-kb_t) \int_{-\infty}^{\infty} \int_{-\infty}^{\infty} \int_{-\infty}^{\infty} \int_{-\infty}^{\infty} dx_1 dy_1 dp_{x1} dp_{y1}$$

$$\times a_{x0}^{\pm}(x_1, y_1, z_1 = 0) \exp\left[i2\pi(p_{x1}x_1 + p_{y1}y_1)\right]$$

$$\times \exp\left[\pi^2 w_0^2 \left(p_{x1}^2 + p_{y1}^2\right) \frac{b_t}{b}\right] \exp\left[-i2\pi(p_{x1}x + p_{y1}y)\right]$$

$$\times \zeta_1^{-1} \exp[i\zeta_1(|z| - ib_t)] \tag{14}$$

Equation (14) is the exact vector potential associated with the extended full-wave generalization of the paraxial Gaussian beam whose input distribution is given by $a_{x0}^{\pm}(x,y,0)$.

2 Cross-spectral density

The electromagnetic fields $E_x(x,y,z)$ and $H_y(x,y,z)$, required for the determination of the z component of the Poynting vector, are obtained by the use of Eqs. (D30) and (D33), respectively, as follows:

$$E_x^{\pm}(x,y,z) = k \exp(-kb_t) \int_{-\infty}^{\infty} \int_{-\infty}^{\infty} \int_{-\infty}^{\infty} \int_{-\infty}^{\infty} dx_1 dy_1 dp_{x1} dp_{y1}$$

$$\times a_{x0}^{\pm}(x_1, y_1, z_1 = 0) \exp\left[i2\pi(p_{x1}x_1 + p_{y1}y_1)\right]$$

$$\times \exp\left[\pi^2 w_0^2 \left(p_{x1}^2 + p_{y1}^2\right) \frac{b_t}{b} ik \left(1 - \frac{4\pi^2 p_{x1}^2}{k^2}\right)\right]$$

$$\times \exp\left[-i2\pi(p_{x1}x + p_{y1}y)\right] \zeta_1^{-1} \exp[i\zeta_1(|z| - ib_t)] \tag{15}$$

and

$$H_y^{\pm}(x,y,z) = \pm k \exp(-kb_t) \int_{-\infty}^{\infty} \int_{-\infty}^{\infty} \int_{-\infty}^{\infty} \int_{-\infty}^{\infty} dx_2 dy_2 dp_{x2} dp_{y2}$$

$$\times a_{x0}^{\pm}(x_2, y_2, z_2 = 0) \exp\left[i2\pi(p_{x2}x_2 + p_{y2}y_2)\right]$$

$$\times \exp\left[\pi^2 w_0^2 \left(p_{x2}^2 + p_{y2}^2\right) \frac{b_t}{b}\right] \exp\left[-i2\pi(p_{x2}x + p_{y2}y)\right]$$

$$\times i \exp[i\zeta_2(|z| - ib_t)] \tag{16}$$

For the extended full Gaussian wave treated in Chapter 6, the electromagnetic fields are fully coherent and there are no random variations of the amplitude over time. The input value of the paraxial approximation of the vector potential, $a_{x0}^{\pm}(x,y,z = 0)$, is known. The integrals in Eqs. (15) and (16) can be evaluated. In general, there are fluctuations, the source electric current density is a random function, and it cannot be specified. Some features of the current source are known. The wave period is T_w, and the time scale of the random variation of the amplitude of the current is T_f. It is known that $T_w \ll T_f$. For example, for light waves, the wave period is on the order of a femtosecond, whereas the time scale of the fluctuations is of the order of a few picoseconds. Therefore, the fluctuating quantities are approximately constants over a wave period. By time averaging over a wave period, the Poynting vector $S_z(\mathbf{r}, t)$ in the $+z$ direction is found from Eqs. (15) and (16) as

$$
\begin{aligned}
S_z(\mathbf{r}, t) = \frac{ck^2}{2} \exp(-2kb_t) \mathrm{Re} \int_{-\infty}^{\infty} & \int_{-\infty}^{\infty} \int_{-\infty}^{\infty} \int_{-\infty}^{\infty} \int_{-\infty}^{\infty} \int_{-\infty}^{\infty} \int_{-\infty}^{\infty} \int_{-\infty}^{\infty} dx_1 dy_1 \\
& \times dp_{x1} dp_{y1} dx_2 dy_2 dp_{x2} dp_{y2} a_{x0}^{\pm}(x_1, y_1, z_1 = 0) a_{x0}^{\pm*}(x_2, y_2, z_2 = 0) \\
& \times \exp[i2\pi(p_{x1}x_1 + p_{y1}y_1)] \exp\left[\pi^2 w_0^2 (p_{x1}^2 + p_{y1}^2) \frac{b_t}{b}\right] \\
& \times ik\left(1 - \frac{4\pi^2 p_{x1}^2}{k^2}\right) \exp[-i2\pi(p_{x1}x + p_{y1}y)] \\
& \times \zeta_1^{-1} \exp[i\zeta_1(|z| - ib_t)] \\
& \times \exp[-i2\pi(p_{x2}x_2 + p_{y2}y_2)] \exp\left[\pi^2 w_0^2 (p_{x2}^2 + p_{y2}^2) \frac{b_t}{b}\right] \\
& \times \exp[i2\pi(p_{x2}x + p_{y2}y)](-i) \exp[-i\zeta_2^*(|z| + ib_t)] \qquad (17)
\end{aligned}
$$

where c is the free space electromagnetic wave velocity, Re denotes the real part, and the asterisk indicates complex conjugation. The slow variations in time caused by the fluctuations are present in $S_z(\mathbf{r}, t)$.

The weakly fluctuating source generates an electromagnetic wave that is spatially localized around the z axis. The detector used for observing the propagation characteristics of the electromagnetic wave has a response time T_d, where $T_f \ll T_d$ [8]. For example, for light waves, the resolution time of the detector used for observation is on the order of a nanosecond. For the detector, the random variations of the Poynting vector averaged over the wave period are too rapid for measurement and only the long time average taken over the response time of the detector can be observed. Instead of averaging $S_z(\mathbf{r}, t)$ over the long response time of the detector, the average is taken over the ensemble. For this purpose, the fluctuations are characterized by a statistically stationary ensemble $\{a_x^{\pm}(x, y, z = 0)\}$, where $a_x^{\pm}(x, y, z = 0)$ is the value of the paraxial approximation of the vector potential at the source plane. An ensemble average of Eq. (17) yields

$$
\begin{aligned}
S_z(\mathbf{r}) = \frac{ck^2}{2} \exp(-2kb_t) \mathrm{Re} \int_{-\infty}^{\infty} & \int_{-\infty}^{\infty} \int_{-\infty}^{\infty} \int_{-\infty}^{\infty} \int_{-\infty}^{\infty} \int_{-\infty}^{\infty} \int_{-\infty}^{\infty} \int_{-\infty}^{\infty} dx_1 dy_1 \\
& \times dp_{x1} dp_{y1} dx_2 dy_2 dp_{x2} dp_{y2} C_0(x_1, y_1; x_2, y_2) \\
& \times \exp[i2\pi(p_{x1}x_1 + p_{y1}y_1)] \exp\left[\pi^2 w_0^2 (p_{x1}^2 + p_{y1}^2) \frac{b_t}{b}\right] \\
& \times ik\left(1 - \frac{4\pi^2 p_{x1}^2}{k^2}\right) \exp[-i2\pi(p_{x1}x + p_{y1}y)]
\end{aligned}
$$

$$\times \zeta_1^{-1} \exp[i\zeta_1(|z| - ib_t)]$$

$$\times \exp[-i2\pi(p_{x2}x_2 + p_{y2}y_2)] \exp\left[\pi^2 w_0^2 \left(p_{x2}^2 + p_{y2}^2\right)\frac{b_t}{b}\right]$$

$$\times \exp[i2\pi(p_{x2}x + p_{y2}y)](-i) \exp[-i\zeta_2^*(|z| + ib_t)] \tag{18}$$

where

$$C_0(x_1, y_1; x_2, y_2) = \langle a_{x0}^{\pm}(x_1, y_1, z_1 = 0)\, a_{x0}^{\pm*}(x_2, y_2, z_2 = 0)\rangle \tag{19}$$

The averaging over the ensemble is denoted by the angle brackets. $C_0(x_1, y_1; x_2, y_2)$ is the cross-spectral density of the fluctuating vector potential across the input plane $z = 0$. The subscript 0 on C denotes the $z = 0$ plane.

From $S_z(\mathbf{r})$, given by Eq. (18), the radiation intensity distribution and the total radiated power are evaluated. These derived quantities are related to the cross-spectral density of the fluctuating vector potential across the input plane $z = 0$. The observations of the radiation intensity distribution and the total radiated power are the physically valid input quantities [9]. The solution of the inverse problem leads to the physically valid cross-spectral density of the fluctuating vector potential across the source plane. An approximate procedure for solving the inverse problem is to obtain the solution of the direct problem for several assumed cross-spectral densities of the fluctuating vector potential across the source plane, and identify the input cross-spectral density that reproduces the observed radiation intensity distribution and the total power. The cross-spectral density of the fluctuating vector potential across the input plane thus determined is physically valid. Here, a start is made toward the solution of the inverse problem by solving completely the direct problem for one assumed cross-spectral density of the vector potential across the source plane by the use of the extended full Gaussian wave governed by Maxwell's equations.

For the fundamental Gaussian beam generated by a fluctuating planar current source, the Schell model [11,12] for the cross-spectral density is of the form

$$C_{0,w}(x_1, y_1; x_2, y_2) = N_w^2 \exp\left(-\frac{x_1^2 + y_1^2}{w_0^2}\right) \exp\left(-\frac{x_2^2 + y_2^2}{w_0^2}\right) g_w(x_1, y_1; x_2, y_2) \tag{20}$$

where N_w is a normalization constant. The complex degree of spatial coherence $g_w(x_1, y_1; x_2, y_2)$ is given by

$$g_w(x_1, y_1; x_2, y_2) = \exp\left\{-\frac{\left[(x_1 - x_2)^2 + (y_1 - y_2)^2\right]}{\sigma_g^2}\right\} \tag{21}$$

where σ_g is the coherence length. For $x_1 - x_2 \neq 0$ and $y_1 - y_2 \neq 0$, as well as for $x_1 = x_2$ and $y_1 = y_2$, $g_w(x_1, y_1; x_2, y_2) = 1$ for $\sigma_g = \infty$. For a fully coherent beam, $\sigma_g = \infty$ and $g_w(x_1, y_1; x_2, y_2) = 1$. Then, the planar source launches the full-wave generalization of a fully coherent fundamental Gaussian beam that has a beam waist w_0 at the input plane. The cross-spectral density of the fully coherent (*fc*) fundamental Gaussian beam is given by

$$C_{0,fc}(x_1, y_1; x_2, y_2) = N_{fc}^2 \exp\left(-\frac{x_1^2 + y_1^2}{w_0^2}\right) \exp\left(-\frac{x_2^2 + y_2^2}{w_0^2}\right) \tag{22}$$

where the subscript *fc* stands for fully coherent. For $\sigma_g \neq \infty$, that is, for finite and nonzero values of σ_g, Eqs. (20) and (21) give the Schell-model cross-spectral density for a partially coherent beam.

For the fundamental Gaussian beam launched by a fluctuating planar current source, the cross-spectral density of the incoherent beam is given by Goodman [13] as

$$C_{0,in}(x_1,y_1;x_2,y_2) = N_{in}^2 \exp\left(-\frac{x_1^2+y_1^2}{w_0^2}\right) \exp\left(-\frac{x_2^2+y_2^2}{w_0^2}\right) \delta(x_1-x_2)\delta(y_1-y_2) \quad (23)$$

where the subscript *in* stands for incoherent. For $x_1 - x_2 \neq 0$ and $y_1 - y_2 \neq 0$, $g_w(x_1,y_1;x_2,y_2) = 0$ for $\sigma_g = 0$. For $x_1 - x_2 = 0$ and $y_1 - y_2 = 0$, $g_w(x_1,y_1;x_2,y_2) = 1$ for $\sigma_g = 0$. Also,

$$\int_{-\infty}^{\infty} \int_{-\infty}^{\infty} g_w(x_1,y_1;x_2,y_2)\, d(x_1-x_2)\, d(y_1-y_2) = \pi\sigma_g^2 = 0$$

for $\sigma_g = 0$. Therefore, $C_{0,w}(x_1,y_1;x_2,y_2)$ given by Eqs. (20) and (21) does not reduce to Eq. (23) as $\sigma_g \to 0$. Hence, $C_{0,w}(x_1,y_1;x_2,y_2)$ cannot be used to represent the cross-spectral density of the fully incoherent fundamental Gaussian beam. For the source current density with weak (w) fluctuations that generate the fully coherent ($\sigma_g = \infty$) and the nearly fully coherent or the partially coherent (σ_g finite and not zero) beams, $C_{0,w}(x_1,y_1;x_2,y_2)$ given by Eqs. (20) and (21) can be used to represent the cross-spectral density of the fluctuating current source.

For the fundamental Gaussian beam generated by a fluctuating planar current source, consider the Schell model for the cross-spectral density modified as follows:

$$C_{0,s}(x_1,y_1;x_2,y_2) = N_s^2 \exp\left(-\frac{x_1^2+y_1^2}{w_0^2}\right) \exp\left(-\frac{x_2^2+y_2^2}{w_0^2}\right) g_s(x_1,y_1;x_2,y_2) \quad (24)$$

where N_s is another normalization constant. The modified complex degree of spatial coherence $g_s(x_1,y_1;x_2,y_2)$ is given by

$$g_s(x_1,y_1;x_2,y_2) = \frac{1}{\pi\sigma_g^2} \exp\left\{-\frac{\left[(x_1-x_2)^2+(y_1-y_2)^2\right]}{\sigma_g^2}\right\} \quad (25)$$

For $x_1 - x_2 \neq 0$ and $y_1 - y_2 \neq 0$, $g_s(x_1,y_1;x_2,y_2) = 0$ for $\sigma_g \to 0$. Also, for $x_1 = x_2$ and $y_1 = y_2$, $g_s(x_1,y_1;x_2,y_2) = \infty$ as $\sigma_g \to 0$. Also,

$$\int_{-\infty}^{\infty} \int_{-\infty}^{\infty} g_s(x_1,y_1;x_2,y_2)\, d(x_1-x_2)\, d(y_1-y_2) = 1$$

and is independent of σ_g. Therefore, $C_{0,s}(x_1,y_1;x_2,y_2)$ given by Eqs. (24) and (25) reduces to Eq. (23) as $\sigma_g \to 0$. Hence, $C_{0,s}(x_1,y_1;x_2,y_2)$ given by Eqs. (24) and (25) can be used to represent the cross-spectral density of the limiting case of the fully incoherent fundamental Gaussian beam.

For $x_1 - x_2 \neq 0$ and $y_1 - y_2 \neq 0$, as well as for $x_1 = x_2$ and $y_1 = y_2$, $g_s(x_1,y_1;x_2,y_2) = 0$ as $\sigma_g \to \infty$. Hence, $C_{0,s}(x_1,y_1;x_2,y_2)$ cannot be used to represent the cross-spectral density of the limiting case of the fully coherent fundamental Gaussian beam. For the source current density with strong (s) fluctuations that generates the fully incoherent $(\sigma_g = 0)$ and the nearly fully incoherent or the partially incoherent (σ_g is zero and finite but not infinite) beams, $C_{0,s}(x_1, y_1; x_2, y_2)$ given by Eqs. (24) and (25) can be used to represent the cross-spectral density of the fluctuating current source.

First, we shall treat the fully coherent and the nearly fully coherent or partially coherent fundamental Gaussian beams. The common (c) and the difference (d) coordinates as defined by

$$u_c = \frac{1}{2}(u_1 + u_2) \quad \text{and} \quad u_d = u_1 - u_2 \quad \text{for } u = x, y \tag{26}$$

are introduced. Then, Eq. (19) is transformed as

$$W_0(x_d, y_d; x_c, y_c) = C_0\left(x_c + \frac{1}{2}x_d, y_c + \frac{1}{2}y_d, x_c - \frac{1}{2}x_d, y_c - \frac{1}{2}y_d\right) \tag{27}$$

For the full-wave generalization of the fundamental Gaussian beam generated by the fluctuating vector potential, the cross-spectral density in the paraxial limit is assumed to be of the Schell-model type [11,12], as given by Eqs. (20) and (21). Expressed in terms of the common and the difference coordinates, the Schell-model cross-spectral density is given by [10]

$$W_{0,w}(x_d, y_d; x_c, y_c) = N_w^2 \exp\left[-\frac{2(x_c^2 + y_c^2)}{w_0^2}\right] \exp\left[-\frac{x_d^2 + y_d^2}{\sigma_t^2}\right] \tag{28}$$

where N_w is a normalization constant to be chosen later in the analysis. Also,

$$\frac{1}{\sigma_t^2} = \frac{1}{2w_0^2} + \frac{1}{\sigma_g^2} \tag{29}$$

where σ_g, the coherence parameter having the dimensions of length, characterizes the spatial coherence properties of the source. For a fully coherent source, $\sigma_g = \infty$, $\sigma_t^2 = 2w_0^2$, and the fundamental Gaussian beam is generated in the paraxial approximation.

The Schell-model sources have been used for the treatment of partially coherent waves because they are known to give intuitively acceptable results for the partially coherent waves [6,7,9]. In addition, for the Schell-model source, the theoretical analysis can be carried out fully and an analytical expression can be deduced for the radiation intensity distribution.

3 Radiation intensity for the partially coherent source

Propagation in $z > 0$ alone is considered since the propagation characteristics in $z < 0$ are obtained from symmetry. The power transported in the $+z$ direction is found from Eq. (18) and (27) as

$$P^+ = \int_{-\infty}^{\infty} \int_{-\infty}^{\infty} dx\, dy\, S_z(\mathbf{r}) \tag{30}$$

$S_z(\mathbf{r})$ from Eq. (18) is substituted into Eq. (30) and the integrations with respect to x and y are performed, yielding $\delta(p_{x1} - p_{x2})$ and $\delta(p_{y1} - p_{y2})$, respectively. Then, the integrations with respect to p_{x2} and p_{y2} are carried out. The result is

$$P^+ = \frac{ck^3}{2} \exp(-2kb_t) \mathrm{Re} \int_{-\infty}^{\infty} \int_{-\infty}^{\infty} \int_{-\infty}^{\infty} \int_{-\infty}^{\infty} \int_{-\infty}^{\infty} \int_{-\infty}^{\infty} dx_1 dy_1$$

$$\times \, dx_2 dy_2 dp_{x1} dp_{y1} W_{0,w}(x_d, y_d; x_c, y_c)$$

$$\times \exp[i2\pi(p_{x1}x_d + p_{y1}y_d)] \exp\left[2\pi^2 w_0^2 (p_{x1}^2 + p_{y1}^2) \frac{b_t}{b}\right]$$

$$\times \left(1 - \frac{4\pi^2 p_{x1}^2}{k^2}\right) \zeta_1^{-1} \exp[iz(\zeta_1 - \zeta_1^*) + b_t(\zeta_1 + \zeta_1^*)] \tag{31}$$

When Eq. (28) is substituted into Eq. (31), in the integrand, x_1 and x_2 appear only in the combination of x_d and x_c. Therefore, the variables are changed from x_1 and x_2 to x_d and x_c. The Jacobian of the transformation is unity. An exactly similar situation exists for the (y_1, y_2) integrations. The x_1 and x_2 integrations are given by

$$I_{x1x2} = \int_{-\infty}^{\infty} dx_c \exp\left(-\frac{2x_c^2}{w_0^2}\right) \int_{-\infty}^{\infty} dx_d \exp\left(-\frac{x_d^2}{\sigma_t^2}\right) \exp(i2\pi p_{x1}x_d) \tag{32}$$

$$= \frac{\pi}{\sqrt{2}} w_0 \sigma_t \exp(-\pi^2 \sigma_t^2 p_{x1}^2) \tag{33}$$

The integrals in Eq. (32) are evaluated, and the result is stated in Eq. (33). Similarly, the y_1 and y_2 integrations are performed, yielding

$$I_{y1y2} = \frac{\pi}{\sqrt{2}} w_0 \sigma_t \exp\left(-\pi^2 \sigma_t^2 p_{y1}^2\right) \tag{34}$$

Substituting Eqs. (28), (33), and (34) into Eq. (31), P^+ is expressed as

$$P^+ = N_w^2 \frac{ck^3}{4} \exp(-2kb_t)\pi^2 w_0^2 \sigma_t^2 \, \mathrm{Re} \int_{-\infty}^{\infty} \int_{-\infty}^{\infty} dp_{x1} dp_{y1}$$

$$\times \exp\left[-2\pi^2 w_0^2 \left(w_\sigma^2 - \frac{b_t}{b}\right)(p_{x1}^2 + p_{y1}^2)\right]\left(1 - \frac{4\pi^2 p_{x1}^2}{k^2}\right)$$

$$\times \zeta_1^{-1} \exp\left[iz(\zeta_1 - \zeta_1^*) + b_t(\zeta_1 + \zeta_1^*)\right] \tag{35}$$

where

$$w_\sigma^2 = \left(1 + \frac{2w_0^2}{\sigma_g^2}\right)^{-1} \tag{36}$$

For a fully coherent beam, $\sigma_g = \infty$, the second term inside the parenthesis in Eq. (36) is zero compared to 1, the first term. For a fully incoherent beam, the second term in Eq. (36) is infinite compared to 1, the first term. As σ_g is reduced from ∞ to $\sigma_g = \sqrt{2}w_0$, the second term is less than or equal to the first term; that is, $\infty \geq \sigma_g^2/2w_0^2 \geq 1$. This range includes the fully coherent beam. Similarly, as σ_g is increased from 0 to $\sigma_g = \sqrt{2}w_0$, the second term is greater than or equal to the first term; that is, $1 \geq \sigma_g^2/2w_0^2 \geq 0$. This range includes the fully incoherent beam. One possible way to separate the nearly coherent and the nearly incoherent beams is to treat the range $\infty \geq \sigma_g^2/2w_0^2 \geq 1$ as defining the partially coherent beams and the range $1 \geq \sigma_g^2/2w_0^2 \geq 0$ as defining the partially incoherent beams. Therefore, we shall use the Schell-model cross-spectral density as given by Eq. (28) to describe the partially coherent beams defined by the range $\infty \geq \sigma_g^2/2w_0^2 \geq 1$. The corresponding range of w_σ for partially coherent beams is $1 \geq w_\sigma \geq 0.7$, where $w_\sigma = 1$ corresponds to the fully coherent beam.

Changing the variables of integration as $2\pi p_{x1} = p\cos\phi$ and $2\pi p_{y1} = p\sin\phi$, Eq. (35) is transformed as

$$P^+ = N_w^2 \frac{ck^2}{16} w_0^2\sigma_t^2 \exp(-2kb_t)\text{Re}\int_0^\infty dp\, p$$

$$\times \int_0^{2\pi} d\phi \left(1 - \frac{p^2}{k^2}\cos^2\phi\right)\exp\left[-\frac{w_0^2}{2}\left(w_\sigma^2 - \frac{b_t}{b}\right)p^2\right]$$

$$\times \xi^{-1}\exp[izk(\xi - \xi^*) + kb_t(\xi + \xi^*)] \tag{37}$$

where

$$\xi = \left(1 - \frac{p^2}{k^2}\right)^{1/2} \tag{38}$$

For $k < p < \infty$, ξ is imaginary and the integrand of Eq. (37) is imaginary. Therefore, the real part is zero and no contribution to P^+ arises. Hence, P^+ given by Eq. (37) simplifies as

$$P^+ = N_w^2 \frac{ck^2}{16} w_0^2\sigma_t^2 \int_0^k dp\, p \int_0^{2\pi} d\phi \left(1 - \frac{p^2}{k^2}\cos^2\phi\right)$$

$$\times \exp\left[-\frac{w_0^2}{2}\left(w_\sigma^2 - \frac{b_t}{b}\right)p^2\right]\xi^{-1}\exp[-2kb_t(1 - \xi)] \tag{39}$$

The paraxial limit $p^2/k^2 \ll 1$ is considered first. Then, ξ in the amplitude is approximated by 1 and in the phase by $1 - p^2/2k^2$. The p^2/k^2 term is neglected in comparison with 1 in the amplitude term $(1 - p^2\cos^2\phi/k^2)$. Then, Eq. (39) simplifies as

$$P_0^+ = N_w^2 \frac{ck^2}{16} w_0^2\sigma_t^2 \int_0^k dp\, p \int_0^{2\pi} d\phi \exp\left(-\frac{1}{4}\sigma_t^2 p^2\right) \tag{40}$$

where the subscript 0 on P^+ stands for paraxial. Since $k^2 \gg p^2$ in the paraxial limit, the upper limit of the p integration is replaced by ∞. Then, the integrals in Eq. (40) are evaluated to yield

$$P_0^+ = N_w^2 \frac{ck^2}{4}\pi w_0^2 \tag{41}$$

As expected, in the paraxial limit, the radiated power is independent of the fluctuations of the vector potential [15,8]. The normalization constant is chosen as

$$N_w = \left(\frac{4}{c\pi w_0^2 k^2} \right)^{1/2} \tag{42}$$

For this choice of N_w, the power transported by the paraxial beam in the $+z$ direction is given by $P_0^+ = 1$ W.

For the chosen value of N_w, Eq. (39) is transformed as

$$P^+ = \frac{\sigma_t^2}{4\pi} \int_0^k dp\, p \int_0^{2\pi} d\phi \left(1 - \frac{p^2}{k^2} \cos^2\phi \right)$$

$$\times \exp\left[-\frac{w_0^2}{2} \left(w_\sigma^2 - \frac{b_t}{b} \right) p^2 \right] \xi^{-1} \exp[-2kb_t(1-\xi)] \tag{43}$$

Changing the variable p in accordance with $p = k\sin\theta$ enables us to cast Eq. (43) as

$$P^+ = \int_0^{\pi/2} d\theta \sin\theta \int_0^{2\pi} d\phi \Phi(\theta,\phi) \tag{44}$$

where

$$\Phi(\theta,\phi) = \frac{k^2 w_0^2 w_\sigma^2}{2\pi} (1 - \sin^2\theta\cos^2\phi)$$

$$\times \exp\left[-\frac{1}{2} k^2 w_0^2 \left(w_\sigma^2 - \frac{b_t}{b} \right) \sin^2\theta \right]$$

$$\times \exp\left[-k^2 w_0^2 \frac{b_t}{b} (1 - \cos\theta) \right] \tag{45}$$

From Eqs. (44) and (45), it is seen that $\Phi(\theta,\phi)$ is the radiation intensity of the partially coherent, extended full-wave generalization of the fundamental Gaussian beam.

For the fully coherent, extended full Gaussian wave, $w_\sigma = 1$, and the radiation intensity given by Eq. (45) with $w_\sigma = 1$ agrees with the expression for the radiation intensity of the extended full Gaussian wave, as given by Eq. (6.22). The radiation intensity $\Phi(\theta,\phi)$ and the time-averaged power P^+ depend on three parameters: b_t/b, kw_0, and w_σ. A variety of characteristics for $\Phi(\theta,\phi)$ and P^+ is possible depending on the values of these parameters. The radiation characteristics are presented only for a select variation of these parameters. The fluctuation parameter w_σ is important. As the fluctuations increase, w_σ decreases from 1, corresponding to a fully coherent wave, to 0, corresponding to a fully incoherent wave. We shall consider the extreme value of $b_t/b = 1$, corresponding to the basic full Gaussian wave. In Fig. 1, for $\phi = 90°$ and $kw_0 = 1.563$, the radiation intensity of the basic full Gaussian wave ($b_t/b = 1$) is depicted as a function of θ for $0° < \theta < 90°$ and for (a) $w_\sigma = 1$, (b) $w_\sigma = 0.9$, (c) $w_\sigma = 0.8$, and (d) $w_\sigma = 0.7$. Figure 1 shows that as the fluctuations increase (that is, as the coherence length decreases), the peak intensity decreases and the localized wave becomes broader. In general, as the fluctuations increase, the peak intensity becomes

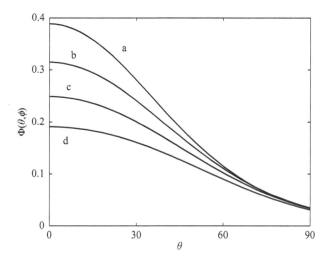

Fig. 1. Radiation intensity $\Phi(\theta, \phi)$ of the partially coherent, extended full Gaussian wave as a function of θ for $0° < \theta < 90°$. The other parameters are $\phi = 90°$, $kw_0 = 1.563$, $b_t/b = 1$, (a) $w_\sigma = 1$, (b) $w_\sigma = 0.9$, (c) $w_\sigma = 0.8$, and (d) $w_\sigma = 0.7$. The normalization is such that the power carried by the corresponding paraxial beam in the $+z$ direction is given by $P_0^+ = 1$ W.

lower and the width of the intensity distribution becomes larger. The treatment of the other limiting case of $b_t/b = 0$, corresponding to the partially coherent fundamental Gaussian wave, is given in [10]. The results for the partially coherent fundamental Gaussian wave ($b_t/b = 0$) are similar to those for the partially coherent basic full Gaussian wave.

4 Time-averaged power for the partially coherent source

From Eqs. (44) and (45), the total power P^+ transported in the $+z$ direction is evaluated. P^+ is a function of kw_0, b_t/b, and w_σ. In Fig. 2, the power P^+ transported in the $+z$ direction is shown as a function of kw_0 for $1 < kw_0 < 5$, for $b_t/b = 1$, and for $w_\sigma = 1, 0.9, 0.8$, and 0.7. The normalization is such that the power carried by the corresponding paraxial ($kw_0 \rightarrow \infty$) beam in the $+z$ direction is given by $P_0^+ = 1$ W. Figure 2 shows that the power P^+ transported in the $+z$ direction increases as kw_0 is increased, approaching the value of $P_0^+ = 1$ W, corresponding to the paraxial beam, as kw_0 becomes large. For kw_0 sufficiently small and well below the paraxial region, the power P^+ decreases as the strength of the fluctuations increases, that is, as the coherence length decreases. In [10], the corresponding result is shown for $b_t/b = 0$. Thus, in general, the power P^+ increases as kw_0 increases, approaching the value of $P_0^+ = 1$ W, corresponding to the paraxial beam, as kw_0 becomes large; for kw_0 below the paraxial region, P^+ decreases as the strength of the fluctuations increases. In Fig. 3, the power P^+ transported in the $+z$ direction is shown as a function of b_t/b for $0 < b_t/b < 1$, for $kw_0 = 1.563$, and for $w_\sigma = 1, 0.9, 0.8$, and 0.7. The normalization is the same as for Fig. 2. It is found from Fig. 3 that the power P^+ transported in the $+z$ direction

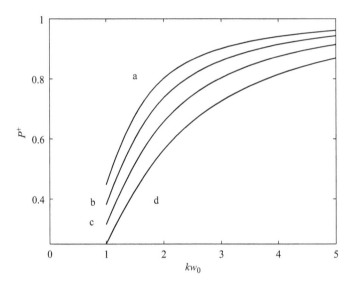

Fig. 2. Power P^+ transported by the partially coherent, extended full Gaussian wave as a function of kw_0 for $1 < kw_0 < 5$. The other parameters are $b_t/b = 1$, (a) $w_\sigma = 1.0$, (b) $w_\sigma = 0.9$, (c) $w_\sigma = 0.8$, and (d) $w_\sigma = 0.7$. The normalization is the same as for Fig. 1.

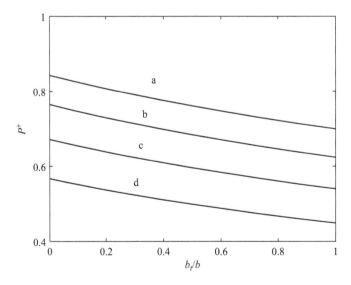

Fig. 3. Power P^+ transported by the partially coherent, extended full Gaussian wave as a function of b_t/b for $0 < b_t/b < 1$. The other parameters are $kw_0 = 1.563$, (a) $w_\sigma = 1.0$, (b) $w_\sigma = 0.9$, (c) $w_\sigma = 0.8$, and (d) $w_\sigma = 0.7$. The normalization is the same as for Fig. 1.

decreases as b_t/b is increased from 0 to 1 for all coherence lengths. For a fixed b_t/b, the power P^+ decreases as the coherence length decreases, or equivalently as w_σ decreases.

5 Radiation intensity for the partially incoherent source

For the fully incoherent source, we shall use the cross-spectral density as given by Eq. (23). Substituting Eq. (23) into Eq. (27) leads to

$$W_{0,\,in}(x_d,y_d;x_c,y_c) = N_{in}^2 \exp\left[-\frac{2(x_c^2 + y_c^2)}{w_0^2}\right]\exp\left[-\frac{(x_d^2 + y_d^2)}{2w_0^2}\right]\delta(x_d)\delta(y_d) \qquad (46)$$

Equation (46) is substituted into Eq. (31). In the integrand, x_1 and x_2 appear only in the combination of x_d and x_c. The variables are changed from x_1 and x_2 to x_d and x_c. A similar situation exists for the (y_1,y_2) integrations. The x_d and x_c integrations, as well as y_d and y_c integrations, are carried out. Then, Eq. (31) simplifies as

$$P^+ = N_{in}^2 \frac{ck^3}{4}\exp(-2kb_t)\pi w_0^2 \operatorname{Re}\int_{-\infty}^{\infty}\int_{-\infty}^{\infty} dp_{x1}dp_{y1}$$
$$\times \exp\left[2\pi^2 w_0^2 \frac{b_t}{b}\left(p_{x1}^2 + p_{y1}^2\right)\right]\left(1 - \frac{4\pi^2 p_{x1}^2}{k^2}\right)$$
$$\times \zeta_1^{-1}\exp\left[iz(\zeta_1 - \zeta_1^*) + b_t(\zeta_1 + \zeta_1^*)\right] \qquad (47)$$

Introducing new integration variables as defined by $2\pi p_{x1} = p\cos\phi$ and $2\pi p_{y1} = p\sin\phi$, Eq. (47) is changed as

$$P^+ = N_{in}^2 \frac{ck^2}{16}\frac{w_0^2}{\pi}\exp(-2kb_t) \operatorname{Re}\int_0^{\infty} dpp\int_0^{2\pi} d\phi\left(1 - \frac{p^2}{k^2}\cos^2\phi\right)\exp\left(\frac{w_0^2}{2}\frac{b_t}{b}p^2\right)$$
$$\times \xi^{-1}\exp[izk(\xi - \xi^*) + kb_t(\xi + \xi^*)] \qquad (48)$$

where ξ is given by Eq. (38). As previously, there is no contribution to P^+ for p in the range $k < p < \infty$. Setting $p = k\sin\theta$ transforms Eq. (48) as

$$P^+ = N_{in}^2 \frac{ck^2}{8}k^2 w_0^2 \frac{1}{2\pi}\int_0^{\pi/2} d\theta\sin\theta\int_0^{2\pi} d\phi(1 - \sin^2\theta\cos^2\phi)$$
$$\times \exp\left(\frac{k^2 w_0^2}{2}\frac{b_t}{b}\sin^2\theta\right)\exp\left[-k^2 w_0^2 \frac{b_t}{b}(1 - \cos\theta)\right] \qquad (49)$$

Equation (49) gives the time-averaged power transported in the $+z$ direction by the fully incoherent extended full Gaussian wave.

For the nearly fully incoherent source, we shall use the modified Schell-model source, as given by Eqs. (24) and (25). The modification consists of changing N_w^2 in the original Shell-model source, as given by Eqs. (20) and (21), by $N_s^2/\pi\sigma_g^2$. The subscripts w and s are used to indicate weak and strong, respectively. From Eq. (39), the time-averaged power transported by the partially incoherent source is found as

$$P^+ = N_s^2 \frac{ck^2}{16\pi} w_0^2 \frac{\sigma_t^2}{\sigma_g^2} \int_0^k dp\, p \int_0^{2\pi} d\phi \left(1 - \frac{p^2}{k^2}\cos^2\phi\right)$$

$$\times \exp\left[-\frac{w_0^2}{2}\left(w_\sigma^2 - \frac{b_t}{b}\right)p^2\right] \xi^{-1}\exp[-2kb_t(1-\xi)] \tag{50}$$

From Eqs. (29) and (36), it is found that

$$\frac{\sigma_t^2}{\sigma_g^2} = \frac{1}{1 + \sigma_g^2/2w_0^2} = 1 - w_\sigma^2 \tag{51}$$

As before, p is changed as $p = k\sin\theta$. Then, Eq. (50), together with Eq. (51), is transformed as

$$P^+ = N_s^2 \frac{ck^2}{8} k^2 w_0^2 (1 - w_\sigma^2) \frac{1}{2\pi} \int_0^{\pi/2} d\theta \sin\theta \int_0^{2\pi} d\phi (1 - \sin^2\theta\cos^2\phi)$$

$$\times \exp\left[-\frac{k^2 w_0^2}{2}\left(w_\sigma^2 - \frac{b_t}{b}\right)\sin^2\theta\right]\exp\left[-k^2 w_0^2\frac{b_t}{b}(1 - \cos\theta)\right] \tag{52}$$

The time-averaged power P^+ for the partially incoherent modified Schell-model source, as given by Eq. (52) for the limiting case of a fully incoherent source corresponding to $w_\sigma = 0$, is the same as the time-averaged power P^+ for the fully incoherent source as given by Eq. (49). Therefore, as is to be expected, Eq. (52) is valid uniformly for w_σ down to and including $w_\sigma = 0$.

It is convenient to restate Eq. (52) as follows:

$$P^+(w_\sigma) = N_s^2 \frac{ck^2}{8} \int_0^{\pi/2} d\theta \sin\theta \int_0^{2\pi} d\phi \Phi(\theta, \phi; w_\sigma) \tag{53}$$

where

$$\Phi(\theta, \phi; w_\sigma) = \frac{k^2 w_0^2 (1 - w_\sigma^2)}{2\pi}(1 - \sin^2\theta\cos^2\phi)$$

$$\times \exp\left[-\frac{k^2 w_0^2}{2}\left(w_\sigma^2 - \frac{b_t}{b}\right)\sin^2\theta\right]\exp\left[-k^2 w_0^2\frac{b_t}{b}(1 - \cos\theta)\right] \tag{54}$$

For the partially incoherent source with fluctuation parameter w_σ varying from 0 to approximately 0.7, the radiation intensity is given by Eq. (54). For $\phi = 0°$, $kw_0 = 1.563$, $b_t/b = 1$, and for four different values of w_σ, namely 0, 0.2, 0.4, and 0.6, the radiation intensity $\Phi(\theta, \phi)$ is shown in Fig. 4 for θ varying from 0° to 90°. The main component of the electric field is in the x direction. Therefore, $\phi = 0°$ corresponds to the E-plane. Even in the presence of strong fluctuations, the spatial localization around the z axis is maintained. As the strength of the fluctuation decreases (that is, as w_σ increases), the peak of the radiation pattern decreases. The wave is flat-topped for $w_\sigma = 0$ and becomes sharper as w_σ increases, although the peak itself decreases.

For $\phi = 90°$, the radiation intensity $\Phi(\theta, \phi)$ is shown in Fig. 5 for θ varying from 0° to 90°. All the other parameters are the same as in Fig. 4. The principal component of the

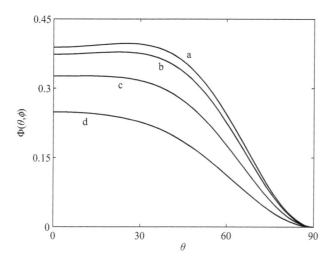

Fig. 4. Radiation intensity $\Phi(\theta, \phi)$ of the partially incoherent, extended full Gaussian wave as a function of θ for $0° < \theta < 90°$. The other parameters are $\phi = 0°$, $kw_0 = 1.563$, $b_t/b = 1$, (a) $w_\sigma = 0$, (b) $w_\sigma = 0.2$, (c) $w_\sigma = 0.4$, and (d) $w_\sigma = 0.6$. The normalization is such that the power carried by the fully incoherent wave in the $+z$ direction is given by $P^+(w_\sigma = 0) = 1$ W.

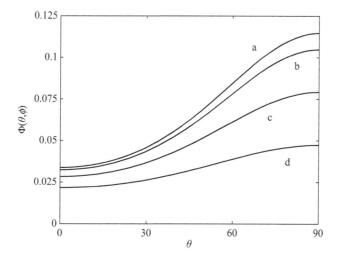

Fig. 5. Radiation intensity $\Phi(\theta, \phi)$ of the partially incoherent, extended full Gaussian wave as a function of θ for $0° < \theta < 90°$. The other parameters are $\phi = 90°$, $kw_0 = 1.563$, $b_t/b = 1$, (a) $w_\sigma = 0$, (b) $w_\sigma = 0.2$, (c) $w_\sigma = 0.4$, and (d) $w_\sigma = 0.6$. The normalization is the same as for Fig. 4.

magnetic field is in the y direction. Therefore, $\phi = 90°$ corresponds to the H-plane. In contrast to the E-plane, in the H-plane, the strong fluctuations destroy the spatial localization around the z axis; that is, $\theta = 0°$. For all values of θ, the intensity appears to decrease as w_σ increases from 0 to 0.6, that is, as the strength of the fluctuations decreases. In Fig. 1, the radiation intensity is shown for the fully coherent and for the nearly fully coherent waves.

All the other parameters in Figs. 1 and 5 are the same. Only the fluctuation parameter w_σ is different. In Fig. 1, the fluctuations are weak and the wave is nearly fully coherent; in Fig. 5, the fluctuations are strong and the wave is nearly fully incoherent. The strong fluctuations eliminate the spatial localization around the z axis ($\theta = 0°$). This result is in some sense to be expected. But in Fig. 5, as the strength of the fluctuations decreases, the intensity decreases in all directions. This result is opposite that in Fig. 1, where as the strength of the fluctuations decreases, the intensity increases in all directions.

6 Time-averaged power for the partially incoherent source

The time-averaged power $P^+(w_\sigma)$ is normalized by choosing N_s^2 such that $P^+(w_\sigma = 0) = 1$ W. Therefore, N_s^2 is defined by

$$1 = N_s^2 \frac{ck^2}{8} \int_0^{\pi/2} d\theta \sin\theta \int_0^{2\pi} d\phi \Phi(\theta, \phi; 0) \tag{55}$$

From Eqs. (53) and (55), the normalized time-averaged power $P^+(w_\sigma)$ is determined as

$$P^+(w_\sigma) = \frac{\int_0^{\pi/2} d\theta \sin\theta \int_0^{2\pi} d\phi \Phi(\theta, \phi; w_\sigma)}{\int_0^{\pi/2} d\theta \sin\theta \int_0^{2\pi} d\phi \Phi(\theta, \phi; 0)} \tag{56}$$

From Eqs. (54) and (56), the total power P^+ transported in the $+z$ direction is shown in Fig. 6 as a function of kw_0 for $1 < kw_0 < 5$, for $b_t/b = 1$, and for $w_\sigma = 0, 0.2, 0.4, 0.6,$ and 0.8. The normalization is the same as that in Fig. 4; that is, the power transported by the fully incoherent ($w_\sigma = 0$) wave in the $+z$ direction is 1 W. Figure 6 shows that for w_σ other than zero, the power P^+ transported in the $+z$ direction decreases as kw_0 is increased. For fixed kw_0 in the range $1 < kw_0 < 5$, the power P^+ transported in the $+z$ direction decreases as w_σ is increased, that is, as the strength of the fluctuations is decreased. These results are in contrast to those for the partially coherent waves shown in Fig. 2. In Fig. 7, the power P^+ transported in the $+z$ direction is shown as a function of b_t/b for $0 < b_t/b < 1$, for $kw_0 = 1.563$, and for $w_\sigma = 0, 0.2, 0.4, 0.6,$ and 0.8. Figure 7 reveals that P^+ is essentially a constant for each w_σ as b_t/b varies over the entire range from 0 to 1. For a fixed b_t/b, the power P^+ decreases as w_σ is increased, that is, as the strength of the fluctuations is decreased. These results are also in contrast with those shown in Fig. 3 for the partially coherent waves.

7 General remarks

For the partially coherent waves, the cross-spectral density is normalized such that the power carried in the $+z$ direction by the corresponding paraxial beam is 1 W. The normalization constant is independent of the fluctuation parameter. For the partially incoherent waves, the power carried in the $+z$ direction by the corresponding paraxial beam is infinite. Therefore, the same normalization procedure cannot be used. Instead, for the partially incoherent

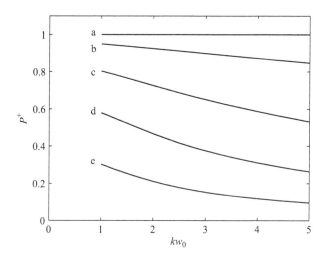

Fig. 6. Power P^+ transported by the partially incoherent, extended full Gaussian wave as a function of kw_0 for $1 < kw_0 < 5$. The other parameters are $b_t/b = 1$, (a) $w_\sigma = 0$, (b) $w_\sigma = 0.2$, (c) $w_\sigma = 0.4$, (d) $w_\sigma = 0.6$, and (e) $w_\sigma = 0.8$. The normalization is the same as for Fig. 4.

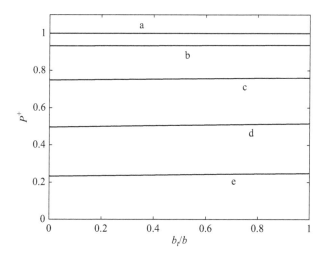

Fig. 7. Power P^+ transported by the partially incoherent, extended full Gaussian wave as a function of b_t/b for $0 < b_t/b < 1$. The other parameters are $kw_0 = 1.563$, (a) $w_\sigma = 0$, (b) $w_\sigma = 0.2$, (c) $w_\sigma = 0.4$, (d) $w_\sigma = 0.6$, and (e) $w_\sigma = 0.8$. The normalization is the same as for Fig. 4.

waves, the normalization is such that the power carried in the $+z$ direction by the corresponding fully incoherent ($w_\sigma = 0$) wave is 1 W.

The radiation intensity and the transported power depend on three parameters: b_t/b, kw_0, and w_σ. A rich variety of propagation characteristics is obtained depending on the values of the three parameters. For the partially coherent waves, the original Schell-model source is used with the coherence length restricted to the range $\infty \geq \sigma_g > \sqrt{2}w_0$. This range could be extended down to but not including $\sigma_g = 0$. Similarly, for the partially incoherent waves, the modified Schell-model source is used, with the coherence length restricted to the range $\sqrt{2}w_0 > \sigma_g \geq 0$. This range could be extended up to but not including $\sigma_g = \infty$. The extensions would increase the available variety of propagation characteristics for the extended full Gaussian waves. If the radiation intensity pattern and the power transported are measured, and if there is a match between the experimentally observed results and some specific predicted results based on the theory developed in this chapter, then an estimate of the cross-spectral density at the source plane can be obtained. The determination of the cross-spectral density at the source plane is the ultimate objective.

References

1. S. R. Seshadri, "Dynamics of the linearly polarized fundamental Gaussian light wave," *J. Opt. Soc. Am. A* **24**, 482–492 (2007).

2. S. R. Seshadri, "Linearly polarized anisotropic Gaussian light wave," *J. Opt. Soc. Am. A* **26**, 1582–1587 (2009).

3. S. R. Seshadri, "Full-wave generalizations of the fundamental Gaussian beam," *J. Opt. Soc. Am. A* **26**, 2515–2520 (2009).

4. S. R. Seshadri, "Reactive power in the full Gaussian light wave," *J. Opt. Soc. Am. A* **26**, 2427–2433 (2009).

5. M. Born and E. Wolf, *Principles of Optics*, 6th ed. (Pergamon, New York, 1984).

6. L. Mandel and E. Wolf, *Optical Coherence and Quantum Optics* (Cambridge University Press, New York, 1995).

7. E. Wolf, *Introduction to the Theory of Coherence and Polarization of Light* (Cambridge University Press, New York, 2007).

8. S. R. Seshadri, "Partially coherent Gaussian Schell-model electromagnetic beams," *J. Opt. Soc. Am. A* **16**, 1373–1380 (1999).

9. A. S. Marathay, *Elements of Optical Coherence Theory* (Wiley, New York, 1982).

10. S. R. Seshadri, "Partially coherent fundamental Gaussian wave generated by a fluctuating planar current source," *J. Opt. Soc. Am. A* **27**, 1372–1377 (2010).

11. A. C. Schell, *The Multiple Plate Antenna* (Doctoral Dissertation, Massachusetts Institute of Technology, 1961), Sec. 7.5.

12. A. C. Schell, "A technique for the determination of the radiation pattern of a partially coherent aperture," *IEEE Trans. Antennas Propag.* AP-**15**, 187–188 (1967).

13. J. W. Goodman, *Statistical Optics* (Wiley, New York, 2000), Sec. 5.5.

14. S. R. Seshadri, "Complex space source theory of partially coherent light wave," *J. Opt. Soc. Am. A* **27**, 1708–1715 (2010).

15. M. W. Kowarz and E. Wolf, "Conservation laws for partially coherent free fields," *J. Opt. Soc. Am. A* **10**, 88–94 (1993).

CHAPTER 15

Airy beams and waves

The Airy functions were analyzed by Kalnins and Miller [1], and the Airy beams were introduced by Berry and Balazs [2] in the context of quantum mechanics. There are other investigations into the properties of the Airy wave packets [3,4]. The original Airy beam cannot be physically realized because infinite power is required to excite the beam. Siviloglou and Christodoulides [5] have introduced a physically realizable Airy beam, and the various properties of this beam have been investigated [6,7]. Bandres and Gutierrez-Vega [8] have analyzed the generalized Airy–Gauss beams that are also physically realizable. All these investigations pertain only to the beams that satisfy the paraxial wave equation. Yan, Yao, Lei, Dan, Yang, and Gao [9], using the virtual source method, have extended the analysis to full Airy waves governed by the exact Helmholtz equation.

In this chapter, some aspects of Airy beams and waves are treated. The fundamental Airy beam and the "finite-energy" (modified) fundamental Airy beam are discussed. The fundamental Airy beam is generalized to obtain the full-wave solution, namely the fundamental Airy wave. For the fundamental Airy wave, the radiation intensity distribution is found to be the same as that for a point electric dipole situated at the origin and oriented normally to the propagation direction. A treatment of the basic full modified Airy wave by the use of the complex space source theory is provided. The propagation characteristics of the basic full modified Airy wave are found to be the same as those for the basic full Gaussian wave, provided that for the former an equivalent waist and an equivalent Rayleigh distance are introduced.

1 Fundamental Airy beam

The secondary source is an infinitesimally thin sheet of electric current located on the $z = 0$ plane. The waves generated by the source propagate in the $+z$ direction in $z > 0$ and in the $-z$ direction in $z < 0$. The source electric current density is in the x direction. The magnetic vector potential is in the x direction and is continuous across the plane $z = 0$. For the generation of a linearly polarized electromagnetic Airy beam, the required x component of the magnetic vector potential at the $z = 0$ plane is specified as

$$a_{x0}^{\pm}(x,y,0) = A_{x0}^{\pm}(x,y,0) = \frac{N}{ik}\mathrm{Ai}(\alpha x)\mathrm{Ai}(\alpha y) \tag{1}$$

where N is a normalization constant, k is the wavenumber, and α is a positive real parameter having the dimension of inverse length. Thus, $1/\alpha$ is the scale length of amplitude variation in the transverse (x, y) plane. The argument of the Airy integrals, $\text{Ai}(\alpha x)$ and $\text{Ai}(\alpha y)$, at the input plane $z = 0$ is real. In Fig. 1, $\text{Ai}(x)$ is shown as a function of x for $-20 < x < 5$. The Airy function decreases monotonically very rapidly for $x > 0$, but decreases in an oscillatory manner very slowly in the $-x$ direction.

The paraxial approximation of the x component of the magnetic vector potential is given by $A_{x0}^{\pm}(x, y, z)$. The additional subscript 0 is used to indicate paraxial. The rapidly varying phase is separated out as in

$$A_{x0}^{\pm}(x, y, z) = \exp(\pm ikz)a_{x0}^{\pm}(x, y, z) \tag{2}$$

The slowly varying amplitude $a_{x0}^{\pm}(x, y, z)$ satisfies the paraxial wave equation:

$$\left(\frac{\partial^2}{\partial x^2} + \frac{\partial^2}{\partial y^2} \pm 2ik\frac{\partial}{\partial z} \right) a_{x0}^{\pm}(x, y, z) = 0 \tag{3}$$

The sign $\pm$ and the superscript $\pm$ are used to indicate that the propagation is in the $\pm z$ direction. The two-dimensional Fourier transform $\bar{a}_{x0}^{\pm}(p_x, p_y, z)$ of $a_{x0}^{\pm}(x, y, z)$ is substituted into Eq. (3). The resulting differential equation is solved to obtain

$$\bar{a}_{x0}^{\pm}(p_x, p_y, z) = \bar{a}_{x0}^{\pm}(p_x, p_y, 0)\exp\left[-\frac{2\pi^2}{k}(p_x^2 + p_y^2)i|z| \right] \tag{4}$$

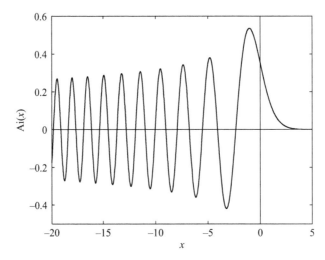

Fig. 1.   Input field $f_0(x, 0) = \text{Ai}(x)$ of the fundamental Airy beam as a function of its argument x for $-20 < x < 5$. For $x > 0$, the field decreases monotonically very rapidly as x increases. The Airy function $\text{Ai}(x)$ has a maximum of 0.5356 for $x = -1.02$. For $x < 0$, the magnitude of the Airy function decreases in an oscillatory manner very slowly as $|x|$ increases.

From Eq. (1), the two-dimensional Fourier transform of $a_{x0}^{\pm}(x, y, 0)$ is found as

$$\bar{a}_{x0}^{\pm}(p_x, p_y, 0) = \frac{N}{ik} \int_{-\infty}^{\infty} dx \exp(i2\pi p_x x) \text{Ai}(\alpha x) \int_{-\infty}^{\infty} dy \exp(i2\pi p_y y) \text{Ai}(\alpha y) \tag{5}$$

Using Eq. (E2), the integrals in Eq. (5) are evaluated to yield

$$\bar{a}_{x0}^{\pm}(p_x, p_y, 0) = \frac{N}{ik} \frac{1}{\alpha^2} \exp\left[-\frac{i}{3}\left(\frac{2\pi}{\alpha}\right)^3 \left(p_x^3 + p_y^3\right)\right] \tag{6}$$

Substituting Eq. (6) into Eq. (4), it is found that

$$\bar{a}_{x0}^{\pm}(p_x, p_y, z) = \frac{N}{ik} \frac{1}{\alpha^2} \exp\left[-\frac{i}{3}\left(\frac{2\pi}{\alpha}\right)^3 \left(p_x^3 + p_y^3\right)\right] \exp\left[-\frac{2\pi^2}{k}\left(p_x^2 + p_y^2\right)i|z|\right] \tag{7}$$

Equation (7) is inverted, with the result that

$$a_{x0}^{\pm}(x, y, z) = \frac{N}{ik} f(x, z) f(y, z) \tag{8}$$

where

$$f(x, z) = \frac{1}{\alpha} \int_{-\infty}^{\infty} dp_x \exp\left[-i2\pi p_x x - \frac{i}{3}\left(\frac{2\pi}{\alpha}\right)^3 p_x^3 - 2\pi^2 i \frac{|z|}{k} p_x^2\right] \tag{9}$$

The integral is evaluated using the method outlined in Appendix E to obtain

$$f(x, z) = \exp\left[i\frac{\alpha^2|z|}{2k}\left(\alpha x - \frac{\alpha^4|z|^2}{6k^2}\right)\right] \text{Ai}\left(\alpha x - \frac{\alpha^4|z|^2}{4k^2}\right) \tag{10}$$

From Eq. (10), $f(x, z)$ and the corresponding expression for $f(y, z)$ are substituted into Eq. (8) and the phase factor is added as stated in Eq. (2) to obtain the x component of the magnetic vector potential as

$$A_{x0}^{\pm}(x, y, z) = \frac{N}{ik} \exp(\pm ikz) \exp\left[i\frac{\alpha^2|z|}{2k}\left(\alpha x + \alpha y - \frac{\alpha^4|z|^2}{3k^2}\right)\right]$$

$$\times \text{Ai}\left(\alpha x - \frac{\alpha^4|z|^2}{4k^2}\right) \text{Ai}\left(\alpha y - \frac{\alpha^4|z|^2}{4k^2}\right) \tag{11}$$

The magnetic vector potential is a product of two Airy integrals at the input plane $z = 0$, and it remains as the product of two Airy integrals as it propagates in the $\pm z$ direction. Therefore, the Airy beam is form invariant on propagation. The maximum of the Airy function occurs when the argument $x = -1.02$. At the input, the peak is close to the z axis, but on propagation, the peak moves away from the z axis in a parabolic fashion in accordance with

$$\alpha x = \alpha y = \frac{\alpha^4 |z|^2}{4k^2} \tag{12}$$

The electric and the magnetic fields are found from the magnetic vector potential by the use of the relations

$$E_{x0}^{\pm}(x,y,z) = \pm H_{y0}^{\pm}(x,y,z) = ikA_{x0}^{\pm}(x,y,z) \tag{13}$$

It is found from Eqs. (11) and (13) that $A_{x0}^{\pm}(x,y,z)$ and $E_{x0}^{\pm}(x,y,z)$ are continuous but $H_{y0}^{\pm}(x,y,z)$ is discontinuous across the source plane $z = 0$. The discontinuity is equivalent to an electric current density induced on the $z = 0$ plane. This electric current density is obtained from Eq. (13) as

$$\mathbf{J}_0(x,y,z) = \hat{z} \times \hat{y}[H_{y0}^{+}(x,y,0) - H_{y0}^{-}(x,y,0)]\delta(z) = -\hat{x}2ikA_{x0}^{+}(x,y,0)\delta(z) \tag{14}$$

Using Eqs. (2) and (14), the complex power is determined as

$$P_C = P_{re} + iP_{im} = -\frac{c}{2}\int_{-\infty}^{\infty} dx \int_{-\infty}^{\infty} dy \int_{-\infty}^{\infty} dz \mathbf{E}_0^{\pm}(x,y,z) \cdot \mathbf{J}_0^*(x,y,z)$$

$$= ck^2 \int_{-\infty}^{\infty} dx \int_{-\infty}^{\infty} dy a_{x0}^{\pm}(x,y,0)a_{x0}^{\pm*}(x,y,0) \tag{15}$$

P_C given by Eq. (15) is real; therefore, $P_{im} = 0$. The reactive power of the Airy beam vanishes.

As previously, Eq. (15) is expressed in terms of the Fourier transform of the slowly varying amplitude $a_{x0}^{\pm}(x,y,0)$ of the vector potential as

$$P_{re} = ck^2 \int_{-\infty}^{\infty} dp_x \int_{-\infty}^{\infty} dp_y \bar{a}_{x0}^{\pm}(p_x,p_y,0)\bar{a}_{x0}^{\pm*}(p_x,p_y,0) \tag{16}$$

Substituting $\bar{a}_{x0}^{\pm}(p_x,p_y,0)$ from Eq. (6) yields

$$P_{re} = \frac{cN^2}{\alpha^4} \int_{-\infty}^{\infty}\int_{-\infty}^{\infty} dp_x dp_y = \infty \tag{17}$$

The time-averaged power associated with the Airy beam is infinite. Therefore, it is not possible to generate the Airy beam in practice. The reason for this behavior is that the Airy integral, for negative values of the argument, does not decrease sufficiently rapidly as the magnitude of the argument becomes very large. This difficulty is overcome by having an exponential function superimposed on the Airy integral; then, the amplitude decays rapidly for negative values of the argument as the magnitude of the argument becomes large.

2 Modified fundamental Airy beam

The fundamental Airy beam cannot be generated in practice because infinite energy is required to sustain the beam. If the fundamental Airy beam is modulated by a function that introduces an exponential decay, the time-averaged power transported by the modified fundamental Airy beam is finite. Such a modified fundamental Airy beam was introduced recently [5]. For the generation of a linearly polarized electromagnetic, modified fundamental Airy beam, the required x component of the magnetic vector potential at the $z = 0$ plane is specified as

$$a_{x0}^{\pm}(x,y,0) = A_{x0}^{\pm}(x,y,0) = \frac{N}{ik}\mathrm{Ai}(\alpha x)\mathrm{Ai}(\alpha y)\exp\left[2\pi a(x+y)\right] \tag{18}$$

where a is real and positive. In Fig. 2, $\mathrm{Ai}(x)\exp(2\pi ax)$ is depicted as a function of x for $-20 < x < 5$ and $a = 0.015$. For $x > 0$, the exponential modulation leaves the rapid monotonic decrease of the field unchanged. But for $x < 0$, the rate of decrease of the magnitude of the field in an oscillatory manner increases considerably as $|x|$ increases.

The variations in the x and y directions are assumed to be symmetric. The two-dimensional Fourier transform of $a_{x0}^{\pm}(x,y,0)$ is obtained from Eq. (18) as

$$\bar{a}_{x0}^{\pm}(p_x,p_y,0) = \frac{N}{ik}\int_{-\infty}^{\infty} dx\,\mathrm{Ai}(\alpha x)\exp\left[i2\pi x(p_x - ia)\right]\int_{-\infty}^{\infty} dy\,\mathrm{Ai}(\alpha y)\exp\left[i2\pi y(p_y - ia)\right] \tag{19}$$

By using Eq. (6), $\bar{a}_{x0}^{\pm}(p_x,p_y,0)$ is determined as

$$\bar{a}_{x0}^{\pm}(p_x,p_y,0) = \frac{N}{ik}\frac{1}{\alpha^2}\exp\left\{-\frac{i}{3}\left(\frac{2\pi}{\alpha}\right)^3\left[(p_x - ia)^3 + (p_y - ia)^3\right]\right\} \tag{20}$$

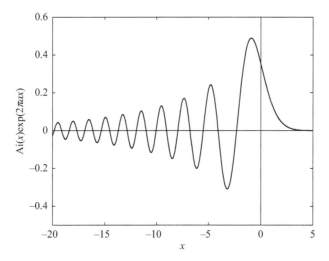

Fig. 2. Input field $f_0(x,0) = \mathrm{Ai}(x)\exp(2\pi ax)$ of the modified fundamental Airy beam as a function of the argument x for $-20 < x < 5$ and $a = 0.015$. For $x > 0$, the exponential modulation does not modify the very rapid monotonic decrease of the field as x increases. For $x < 0$, the exponential modulation significantly changes the field behavior; the rate of decrease of the magnitude of the field in an oscillatory manner increases considerably as $|x|$ increases.

For propagation through a distance z, there is a quadratic phase change, as given by Eq. (4). This phase change is included to obtain $\overline{a}_{x0}^{\pm}(p_x, p_y, z)$ from $\overline{a}_{x0}^{\pm}(p_x, p_y, 0)$ as

$$\overline{a}_{x0}^{\pm}(p_x, p_y, z) = \frac{N}{ik}\overline{f}_0(p_x, z)\overline{f}_0(p_y, z) \tag{21}$$

where

$$\overline{f}_0(p_x, z) = \frac{1}{a}\exp\left\{-\frac{i}{3}\left(\frac{2\pi}{a}\right)^3(p_x - ia)^3 - \frac{2\pi^2}{k}i|z|p_x^2\right\} \tag{22}$$

The inverse Fourier transform of $\overline{f}_0(p_x, z)$ is given by

$$f_0(x, z) = \frac{1}{a}\int_{-\infty}^{\infty} dp_x \exp\left\{-i2\pi p_x x - \frac{i}{3}\left(\frac{2\pi}{a}\right)^3(p_x - ia)^3 - \frac{2\pi^2}{k}i|z|p_x^2\right\} \tag{23}$$

The method explained in Appendix E is used to evaluate the integral. The terms in the exponent are arranged in powers of p_x. Then, the p_x^3 and p_x^2 terms are combined to form a perfect cube of the function:

$$\xi_x = p_x - i\left(a + \frac{a^3 i|z|}{4\pi k}\right) \tag{24}$$

In the remaining terms, p_x is expressed in terms of ξ_x. There are only ξ_x^3, ξ_x, and constant terms with no ξ_x^2 term. Therefore, Eq. (23) is transformed as

$$\begin{aligned}
f_0(x, z) = \frac{1}{a}\int_{-\infty}^{\infty} d\xi_x \exp\Bigg\{&-\frac{i}{3}\left(\frac{2\pi}{a}\right)^3\xi_x^3 - i2\pi\xi_x\left[x + \frac{i|z|}{k}\left(2\pi a + \frac{ia^3|z|}{4k}\right)\right] \\
&+ 2\pi x\left(a + \frac{a^3 i|z|}{4\pi k}\right) + \frac{2\pi^2 a^2 i|z|}{k} - \frac{\pi a a^3|z|^2}{k^2} - \frac{a^6 i|z|^3}{12k^3}\Bigg\}
\end{aligned} \tag{25}$$

Then, Eq. (E2) is used to express $f_0(x, z)$ as Airy integral as follows:

$$\begin{aligned}
f_0(x, z) = \exp\Bigg[&2\pi x\left(a + \frac{a^3 i|z|}{4\pi k}\right) + \frac{2\pi^2 a^2 i|z|}{k} - \frac{\pi a a^3|z|^2}{k^2} - \frac{a^6 i|z|^3}{12k^3} \\
&\times \mathrm{Ai}\left\{a\left[x + \frac{i|z|}{k}\left(2\pi a + \frac{ia^3|z|}{4k}\right)\right]\right\}\Bigg]
\end{aligned} \tag{26}$$

The expression for $f_0(y, z)$ is the same as that for $f_0(x, z)$ with x replaced by y. For $|z| = 0$, Eq. (26) reproduces the assumed input value. For $a = 0$, Eq. (26) reproduces the corresponding value for the fundamental Airy beam as given by Eq. (10). From Eq. (26), $f_0(x, z)$ and the corresponding expression for $f_0(y, z)$ are substituted into Eq. (8) and the plane-wave phase factor is added as given by Eq. (2) to obtain the x component of the magnetic vector potential as

$$A_{x0}^{\pm}(x,y,z) = \frac{N}{ik}\exp(\pm ikz)\exp\left[\left(2\pi a + \frac{a^3 i|z|}{2k}\right)(x+y) + \frac{4\pi^2 a^2 i|z|}{k} - \frac{2\pi a a^3 |z|^2}{k^2} - \frac{a^6 i|z|^3}{6k^3}\right]$$

$$\times \text{Ai}\left\{a\left[x + \frac{i|z|}{k}\left(2\pi a + \frac{ia^3|z|}{4k}\right)\right]\right\}\text{Ai}\left\{a\left[y + \frac{i|z|}{k}\left(2\pi a + \frac{ia^3|z|}{4k}\right)\right]\right\}$$

(27)

For the modified fundamental Airy beam as well, it follows from Eq. (15) that the reactive power vanishes. The real power transmitted by the beam is obtained by substituting $\bar{a}_{x0}^{\pm}(p_x,p_y,0)$ from Eq. (20) into Eq. (16), with the result that

$$P_{re} = \frac{cN^2}{\alpha^4}\int_{-\infty}^{\infty} dp_x \exp\left\{-\frac{i}{3}\left(\frac{2\pi}{\alpha}\right)^3\left[(p_x - ia)^3 - (p_x + ia)^3\right]\right\}$$

$$\times \int_{-\infty}^{\infty} dp_y \exp\left\{-\frac{i}{3}\left(\frac{2\pi}{\alpha}\right)^3\left[(p_y - ia)^3 - (p_y + ia)^3\right]\right\}$$

(28)

Equation (28) is simplified to yield

$$P_{re} = \frac{cN^2}{\alpha^4}\exp\left[\frac{4}{3}\left(\frac{2\pi a}{\alpha}\right)^3\right]\int_{-\infty}^{\infty}\int_{-\infty}^{\infty} dp_x dp_y \exp\left[-\left(\frac{2\pi}{\alpha}\right)^3 2a(p_x^2 + p_y^2)\right]$$

(29)

The power spectrum, in particular the dependence on $(p_x^2 + p_y^2)$, for the modified fundamental Airy beam is similar to that for the Gaussian beam. The integrations in Eq. (29) are completed to find the power transmitted by the paraxial beam as

$$P_{re} = \frac{cN^2}{16\pi^2 \alpha a}\exp\left[\frac{4}{3}\left(\frac{2\pi a}{\alpha}\right)^3\right]$$

(30)

For $a \neq 0$, P_{re} is finite. Thus, as expected, the introduction of an exponential decay for the negative values of the argument x leads to a physically realizable fundamental Airy beam.

3 Fundamental Airy wave

The secondary source for the approximate paraxial beams and the exact full waves is a current sheet that is situated on the plane $z = 0$. The response of the electric current source given by Eq. (14), obtained in the paraxial approximation, is the fundamental Airy beam. The same current source for the full Helmholtz equation yields the fundamental Airy wave. From Eqs. (1) and (14), the electric current density generating the fundamental Airy beam is found as

$$\mathbf{J}_0(x,y,z) = -\hat{x}2N\text{Ai}(\alpha x)\text{Ai}(\alpha y)\delta(z)$$

(31)

The electric current density is in the x direction; therefore, the exact vector potential $A_x^\pm(x, y, z)$ is also in the x direction. From Eqs. (31) and (D25), $A_x^\pm(x, y, z)$ is found to be governed by the following inhomogeneous Helmholtz wave equation:

$$\left(\frac{\partial^2}{\partial x^2} + \frac{\partial^2}{\partial y^2} + \frac{\partial^2}{\partial z^2} + k^2\right)A_x^\pm(x, y, z) = 2N\,\mathrm{Ai}(\alpha x)\mathrm{Ai}(\alpha y)\delta(z) \tag{32}$$

The Fourier integral representation of $\mathrm{Ai}(\alpha x)\mathrm{Ai}(\alpha y)$ is found from Eq. (9) as

$$\mathrm{Ai}(\alpha x)\mathrm{Ai}(\alpha y) = \frac{1}{\alpha^2}\int_{-\infty}^{\infty}\int_{-\infty}^{\infty} dp_x dp_y \exp[-i2\pi(p_x x + p_y y)]\exp\left[-\frac{i}{3}\left(\frac{2\pi}{\alpha}\right)^3\left(p_x^3 + p_y^3\right)\right] \tag{33}$$

Substituting the Fourier integral representation $\overline{A}_x^\pm(p_x, p_y, z)$ of $A_x^\pm(x, y, z)$ and Eq. (33) into Eq. (32), it is obtained that

$$\left(\frac{\partial^2}{\partial z^2} + \zeta^2\right)\overline{A}_x^\pm(p_x, p_y, z) = \frac{2N}{\alpha^2}\delta(z)\exp\left[-\frac{i}{3}\left(\frac{2\pi}{\alpha}\right)^3\left(p_x^3 + p_y^3\right)\right] \tag{34}$$

where

$$\zeta = \left[k^2 - 4\pi^2\left(p_x^2 + p_y^2\right)\right]^{1/2} \tag{35}$$

Equation (34) is solved as discussed in Appendix A.2. The solution is inverted, with the result that

$$A_x^\pm(x, y, z) = -\frac{2N}{\alpha^2}\int_{-\infty}^{\infty}\int_{-\infty}^{\infty} dp_x dp_y \exp[-i2\pi(p_x x + p_y y)]$$
$$\times \exp\left[-\frac{i}{3}\left(\frac{2\pi}{\alpha}\right)^3\left(p_x^3 + p_y^3\right)\right]\frac{i}{2\zeta}\exp(i\zeta|z|) \tag{36}$$

The paraxial approximation corresponds to $4\pi^2(p_x^2 + p_y^2) \ll k^2$. When ζ in Eq. (35) is expanded into a power series in $4\pi^2(p_x^2 + p_y^2)/k^2$, the first two terms are given by Eq. (C17). In Eq. (36), if ζ in the amplitude is replaced by the first term in Eq. (C17) and ζ in the phase is replaced by the first two terms in Eq. (C17), the paraxial approximation to Eq. (36) is found as

$$A_x^\pm(x, y, z) = -\frac{Ni}{\alpha^2 k}\exp(\pm ikz)\int_{-\infty}^{\infty}\int_{-\infty}^{\infty} dp_x dp_y \exp[-i2\pi(p_x x + p_y y)]$$
$$\times \exp\left[-\frac{i}{3}\left(\frac{2\pi}{\alpha}\right)^3(p_x^3 + p_y^3) - i\frac{2\pi^2}{k}(p_x^2 + p_y^2)|z|\right] \tag{37}$$

Using Eq. (9), Eq. (37) is expressed as

$$A_x^\pm(x, y, z) = \frac{N}{ik}\exp(\pm ikz)f(x, z)f(y, z) \tag{38}$$

From Eqs. (2) and (8), it is verified that Eq. (38) represents correctly the fundamental Airy beam. The exact vector potential reduces correctly to the paraxial beam in the appropriate limit.

The x component of the electric field is found from Eqs. (36) and (D30) as

$$
E_x^{\pm}(x,y,z) = \frac{N}{\alpha^2 i} \int_{-\infty}^{\infty} \int_{-\infty}^{\infty} dp_x dp_y \, \exp\left[-i2\pi(p_x x + p_y y)\right]
$$

$$
\times \exp\left[-\frac{i}{3}\left(\frac{2\pi}{\alpha}\right)^3 \left(p_x^3 + p_y^3\right)\right] ik\left(1 - \frac{4\pi^2 p_x^2}{k^2}\right)\frac{1}{\zeta}\exp(i\zeta|z|) \qquad (39)
$$

The complex power is determined from Eq. (D18) as

$$
P_C = -\frac{c}{2}\int_{-\infty}^{\infty}\int_{-\infty}^{\infty}\int_{-\infty}^{\infty} dx\, dy\, dz\, E_x^{\pm}(x,y,z) J_{x0}^*(x,y,z) \qquad (40)
$$

The electric field $E_x^{\pm}(x,y,z)$ is substituted from Eq. (39), the current density $J_{x0}^*(x,y,z)$ is substituted from Eqs. (31) and (33), and the integration with respect to z is performed to obtain

$$
P_C = -\frac{c}{2}\int_{-\infty}^{\infty}\int_{-\infty}^{\infty} dx\, dy\, \frac{N}{\alpha^2 i}\int_{-\infty}^{\infty}\int_{-\infty}^{\infty} dp_x dp_y \exp\left[-i2\pi(p_x x + p_y y)\right]
$$

$$
\times \exp\left[-\frac{i}{3}\left(\frac{2\pi}{\alpha}\right)^3 \left(p_x^3 + p_y^3\right)\right] ik\left(1 - \frac{4\pi^2 p_x^2}{k^2}\right)\frac{1}{\zeta}
$$

$$
\times (-2)\frac{N}{\alpha^2}\int_{-\infty}^{\infty}\int_{-\infty}^{\infty} dp_{x1} dp_{y1} \exp\left[i2\pi(p_{x1} x + p_{y1} y)\right]
$$

$$
\times \exp\left[\frac{i}{3}\left(\frac{2\pi}{\alpha}\right)^3 \left(p_{x1}^3 + p_{y1}^3\right)\right] \qquad (41)
$$

The integrations with respect to x and y are carried out first; the result is the product of two delta functions: $\delta(p_x - p_{x1})\delta(p_y - p_{y1})$. The integrations with respect to p_{x1} and p_{y1} are performed next. Then, P_C given by Eq. (41) simplifies as

$$
P_C = \frac{ckN^2}{\alpha^4}\int_{-\infty}^{\infty}\int_{-\infty}^{\infty} dp_x dp_y \left(1 - \frac{4\pi^2 p_x^2}{k^2}\right)\frac{1}{\zeta} \qquad (42)
$$

The variables of integration are changed as follows: $2\pi p_x = p\cos\phi$ and $2\pi p_y = p\sin\phi$. Then, Eq. (42) is changed as

$$
P_C = \frac{ckN^2}{4\pi^2\alpha^4}\int_0^{\infty} dp\, p \int_0^{2\pi} d\phi \left(1 - \frac{p^2\cos^2\phi}{k^2}\right)(k^2 - p^2)^{-1/2} \qquad (43)
$$

For $k < p < \infty$, setting $p^2 = k^2(1 + \tau^2)$, the value of P_C is given by

$$iP_{im} = -i\frac{ck^2N^2}{4\pi^2\alpha^4}\int_0^\infty d\tau \int_0^{2\pi} d\phi\left[1 - \cos^2\phi\left(1 + \tau^2\right)\right] \tag{44}$$

It is seen from Eq. (44) that the reactive power, the imaginary part of P_C, becomes infinite. The real power is obtained from Eq. (43) as

$$P_{re} = \frac{ckN^2}{4\pi^2\alpha^4}\int_0^k dp\,p \int_0^{2\pi} d\phi\left(1 - \frac{p^2\cos^2\phi}{k^2}\right)(k^2 - p^2)^{-1/2} \tag{45}$$

Let $p = k\sin\theta$. Then, Eq. (45) can be recast as

$$P_{re} = \frac{ck^2N^2}{4\pi^2\alpha^4}\int_0^{\pi/2} d\theta\sin\theta \int_0^{2\pi} d\phi(1 - \sin^2\theta\cos^2\phi) \tag{46}$$

The radiation intensity distribution $(1 - \sin^2\theta\cos^2\phi)$ is the same as that for a point electric dipole. The integrations in Eq. (46) are carried out. The result is

$$P_{re} = \frac{ck^2N^2}{3\pi\alpha^4} \tag{47}$$

The normalization constant N is chosen as

$$N = \frac{\sqrt{6\pi}\alpha^2}{\sqrt{ck}} \tag{48}$$

Then, the real power generated by the fundamental Airy wave is $P_{re} = 2$ W. The time-averaged power transported by the fundamental Airy wave is denoted by $P_{FA}^{\pm}$. By symmetry, $P_{FA}^{+} = P_{FA}^{-}$. One half of the generated real power is transported in the $+z$ direction and the other half is carried in the $-z$ direction. Therefore, $P_{FA}^{+} = P_{FA}^{-} = \frac{1}{2}P_{re} = 1$ W. The normalization constant N is such that the time-averaged power $P_{FA}^{\pm}$ transported by the fundamental Airy wave in the $\pm z$ direction is given by $P_{FA}^{\pm} = 1$ W.

4 Basic full modified Airy wave

Deschamps [10] observed that the paraxial approximation of the field due to a point source reproduced the fundamental Gaussian beam, provided that the location coordinates are in the complex space. The location in the complex space for the point source is given by $x = 0$, $y = 0$, and $|z| = ib$. Here $b = \frac{1}{2}kw_0^2$ is the Rayleigh distance and w_0 is the waist of the Gaussian beam at the input plane. Postulation of the required complex space source for other than the fundamental Gaussian beam is difficult. Such a difficult task was performed by Bandres and Gutierrez-Vega [11] for the full-wave generalization of the higher-order hollow Gaussian beam, and by Zhang, Song, Chen, Ji, and Shi [12] for the full-wave generalization of the cosh-Gauss paraxial beam. These higher-order beams and waves are based on the Gaussian beam, and the source location depends on the Rayleigh distance $b[|z| = ib]$.

The discovered sources are higher-order point sources at the same location or the same point source at different locations in the transverse (x, y) plane. The first change from $|z| = ib$ was made in 2009 [13], and this too is for Gaussian beam–based paraxial beams and full waves. Recently Yan, Yao, Lei, Dan, Yang, and Gao [9] postulated the complex space higher-order point source for the full-wave generalization of the "finite-energy" fundamental Airy beam. For this paraxial beam, there is no *a priori* information on the location of the point source in the complex space. For the finite-energy fundamental Airy beam, the source location in the complex space and the higher-order point sources needed for the full-wave generalization are derived starting from the paraxial beam solution.

From Eqs. (8) and (18), the part of the input field that depends on x is given by

$$f_0(x, 0) = \text{Ai}(\alpha x)\exp(2\pi a x) \tag{49}$$

As before, the subscript 0 is used to denote paraxial. The corresponding part of the paraxial beam that depends on x is obtained in Eq. (23). The cube term in Eq. (23) is expanded and the terms are rearranged, with the result that

$$f_0(x, z) = \frac{1}{\alpha}\exp\left[\frac{1}{3}\left(\frac{2\pi a}{\alpha}\right)^3\right]\int_{-\infty}^{\infty} dp_x \exp\left[-i2\pi p_x x\right.$$
$$\left. -\frac{i}{3}\left(\frac{2\pi}{\alpha}\right)^3 p_x^3 - \frac{2\pi^2}{k}\left(i|z| + \frac{4\pi ka}{\alpha^3}\right)p_x^2 + i\left(\frac{2\pi}{\alpha}\right)^3 a^2 p_x\right] \tag{50}$$

As shown in Eqs. (5.2)–(5.5) for the fundamental Gaussian beam, the source location is chosen to annul the p_x^2 term. Therefore, the source location in the complex space is chosen as

$$|z| = ib = i\frac{4\pi ka}{\alpha^3} \tag{51}$$

Then, $f_0(x, |z| = ib)$ gives the part of the virtual source that depends on x. The constant term is taken outside the integral sign. The p_x^3 and p_x terms are removed by differential operations on the $2\pi p_x x$ term. An exponential differential operator is required. Consider the differential operator consisting of an infinite series of terms:

$$L = \sum_{n=0}^{n=\infty} \frac{g^n}{n!} = \exp(g) \tag{52}$$

The operator acts on the function that appears on its right side. The operator L is written as an exponential operator, as indicated in Eq. (52). Then, $f_0(x, |z| = ib)$ is transformed as

$$f_0(x, |z| = ib) = \frac{1}{\alpha}\exp\left[\frac{1}{3}\left(\frac{2\pi a}{\alpha}\right)^3\right]\exp\left\{\frac{1}{3}\left[-\frac{1}{\alpha^3}\frac{\partial^3}{\partial x^3} - \frac{3}{\alpha}\left(\frac{2\pi a}{\alpha}\right)^2\frac{\partial}{\partial x}\right]\right\}$$
$$\times \int_{-\infty}^{\infty} dp_x \exp(-i2\pi p_x x) \tag{53}$$

The operations are on the function $\exp(-i2\pi p_x x)$. The differential operator is given by

$$D(u) = \frac{1}{3}\left[-\frac{1}{a^3}\frac{\partial^3}{\partial u^3} - \frac{3}{a}\left(\frac{2\pi a}{a}\right)^2 \frac{\partial}{\partial u}\right] \quad \text{for } u = x, y \tag{54}$$

Then, Eq. (53) is expressed concisely as

$$f_0(x, |z| = ib) = \frac{1}{a}\exp\left[\frac{1}{3}\left(\frac{2\pi a}{a}\right)^3\right]\exp[D(x)]\delta(x) \tag{55}$$

Similarly, the part of the virtual source that depends on y is given by

$$f_0(y, |z| = ib) = \frac{1}{a}\exp\left[\frac{1}{3}\left(\frac{2\pi a}{a}\right)^3\right]\exp[D(y)]\delta(y) \tag{56}$$

Then, the virtual source for the vector potential given by Eq. (18) is

$$C_{s0}(x,y) = \frac{N}{ik}f_0(x, |z| = ib)f_0(y, |z| = ib)$$

$$= \frac{N}{ik}\frac{1}{a^2}\exp\left[\frac{2}{3}\left(\frac{2\pi a}{a}\right)^3\right]\exp[D(x) + D(y)]\delta(x)\delta(y) \tag{57}$$

The source given by Eq. (57), located at $|z| = ib$, generates the paraxial beam for $|z| > 0$. The same source at $|z| = 0$ produces the asymptotic limit of the paraxial beam. Because the paraxial wave equation is only an approximation for the full Helmholtz equation, the complex space source deduced from the paraxial beam, together with an excitation coefficient S_{ex}, is used for the Helmholtz equation to obtain the asymptotic value of the full Airy wave. Hence, the source for generating the asymptotic limit for the full-wave solution of the Helmholtz equation is found from Eq. (57) as

$$C_{s0}(x,y) = S_{ex}\frac{N}{ik}f_0(x, |z| = ib)f_0(y, |z| = ib)\delta(z) \tag{58}$$

The basic full-wave generalizations were obtained for a number of paraxial beams. The excitation coefficient was found to be the same. On that basis, the excitation coefficient is assumed as

$$S_{ex} = -2ik\exp(-kb) \tag{59}$$

where b is given by Eq. (51). An equivalent waist is introduced as

$$w_{0,eq} = \left(\frac{8\pi a}{a^3}\right)^{1/2} \tag{60}$$

Then, the equivalent Rayleigh distance is found as

$$b = \frac{1}{2}kw_{0,eq}^2 = \frac{4\pi ka}{a^3} \tag{61}$$

which is the same as that stated in Eq. (51). Therefore, b is referred to as the equivalent Rayleigh distance.

Let $G(x, y, z)$ be the solution of the Helmholtz equation for the source given by Eq. (58). It is convenient to express the source in the form given by Eq. (50). Therefore, the inhomogeneous differential equation satisfied by $G(x, y, z)$ is given by

$$\left(\frac{\partial^2}{\partial x^2} + \frac{\partial^2}{\partial y^2} + \frac{\partial^2}{\partial z^2} + k^2\right) G(x, y, z) = 2ik \exp(-kb)\frac{N}{ik}\delta(z)\frac{1}{a^2}\exp\left[\frac{2}{3}\left(\frac{2\pi a}{a}\right)^3\right]$$

$$\times \int_{-\infty}^{\infty} dp_x \exp(-i2\pi p_x x)\exp\left[-\frac{i}{3}\left(\frac{2\pi}{a}\right)^3 p_x^3 + i\left(\frac{2\pi}{a}\right)^3 a^2 p_x\right]$$

$$\times \int_{-\infty}^{\infty} dp_y \exp(-i2\pi p_y y)\exp\left[-\frac{i}{3}\left(\frac{2\pi}{a}\right)^3 p_y^3 + i\left(\frac{2\pi}{a}\right)^3 a^2 p_y\right] \tag{62}$$

$G(x, y, z)$ is expressed in terms of its two-dimensional Fourier transform $\overline{G}(p_x, p_y, z)$. From Eq. (62), $\overline{G}(p_x, p_y, z)$ is found to satisfy the following differential equation:

$$\left(\frac{\partial^2}{\partial z^2} + \zeta^2\right)\overline{G}(p_x, p_y, z) = 2\exp(-kb)\frac{N}{a^2}\delta(z)\exp\left[\frac{2}{3}\left(\frac{2\pi a}{a}\right)^3\right]$$

$$\times \exp\left[-\frac{i}{3}\left(\frac{2\pi}{a}\right)^3 p_x^3 + i\left(\frac{2\pi}{a}\right)^3 a^2 p_x\right]$$

$$\times \exp\left[-\frac{i}{3}\left(\frac{2\pi}{a}\right)^3 p_y^3 + i\left(\frac{2\pi}{a}\right)^3 a^2 p_y\right] \tag{63}$$

where

$$\zeta = \left[k^2 - 4\pi^2\left(p_x^2 + p_y^2\right)\right]^{1/2} \tag{64}$$

The one-dimensional Green's function is obtained from the solution of Eq. (63) and inverted to yield

$$G(x, y, z) = -i\exp(-kb)\frac{N}{a^2}\exp\left[\frac{2}{3}\left(\frac{2\pi a}{a}\right)^3\right]\int_{-\infty}^{\infty}\int_{-\infty}^{\infty} dp_x dp_y \exp[-i2\pi(p_x x + p_y y)]$$

$$\times \exp\left[-\frac{i}{3}\left(\frac{2\pi}{a}\right)^3 p_x^3 + i\left(\frac{2\pi}{a}\right)^3 a^2 p_x\right]\exp\left[-\frac{i}{3}\left(\frac{2\pi}{a}\right)^3 p_y^3 + i\left(\frac{2\pi}{a}\right)^3 a^2 p_y\right]$$

$$\times \frac{1}{\zeta}\exp(i\zeta|z|) \tag{65}$$

$G(x, y, z)$ is the asymptotic value of the basic full-wave generalization of the finite-energy Airy beam. By analytically continuing $G(x, y, z)$ from $|z|$ to $|z| - ib$, the x component of the magnetic vector potential that generates the basic full modified Airy wave is found as

$$
A_x^{\pm}(x, y, z) = -i \exp(-kb) \frac{N}{\alpha} \int_{-\infty}^{\infty} dp_x \exp(-i2\pi p_x x)
$$

$$
\times \exp\left[-\frac{i}{3}\left(\frac{2\pi}{\alpha}\right)^3 p_x^3 + i\left(\frac{2\pi}{\alpha}\right)^3 a^2 p_x + \frac{1}{3}\left(\frac{2\pi a}{\alpha}\right)^3 \right]
$$

$$
\times \frac{1}{\alpha} \int_{-\infty}^{\infty} dp_y \exp(-i2\pi p_y y)
$$

$$
\times \exp\left[-\frac{i}{3}\left(\frac{2\pi}{\alpha}\right)^3 p_y^3 + i\left(\frac{2\pi}{\alpha}\right)^3 a^2 p_y + \frac{1}{3}\left(\frac{2\pi a}{\alpha}\right)^3 \right]
$$

$$
\times \frac{1}{\zeta} \exp[i\zeta(|z|) - ib] \tag{66}
$$

For the paraxial approximation, ζ given by Eq. (64) is expanded as follows

$$
\zeta = k - \frac{2\pi^2}{k}(p_x^2 + p_y^2) \tag{67}
$$

In Eq. (66), ζ is replaced by the first term in the amplitude and by the first two terms in the phase from Eq. (67). The result is

$$
A_{x0}^{\pm}(x, y, z) = \frac{N}{ik} \exp(\pm ikz) \frac{1}{\alpha} \int_{-\infty}^{\infty} dp_x \exp\left[-i2\pi p_x x - \frac{i}{3}\left(\frac{2\pi}{\alpha}\right)^3 p_x^3 + i\left(\frac{2\pi}{\alpha}\right)^3 a^2 p_x \right.
$$

$$
\left. + \frac{1}{3}\left(\frac{2\pi a}{\alpha}\right)^3 - \frac{2\pi^2}{k}(i|z| + b)p_x^2 \right]
$$

$$
\times \frac{1}{\alpha} \int_{-\infty}^{\infty} dp_y \exp\left[-i2\pi p_y y - \frac{i}{3}\left(\frac{2\pi}{\alpha}\right)^3 p_y^3 + i\left(\frac{2\pi}{\alpha}\right)^3 a^2 p_y \right.
$$

$$
\left. + \frac{1}{3}\left(\frac{2\pi a}{\alpha}\right)^3 - \frac{2\pi^2}{k}(i|z| + b)p_y^2 \right] \tag{68}
$$

A comparison of Eq. (68) with Eqs. (50), (8), and (2) shows that the exact vector potential correctly reproduces the paraxial beam approximation in the appropriate limit.

We shall use the exact vector potential given by Eq. (66) to derive some of the propagation characteristics of the basic full modified Airy wave. The vector potential in the x direction does not generate the magnetic field component in the x direction. Therefore, the only field components that contribute to the Poynting vector in the z direction are $E_x^{\pm}(x, y, z)$

and $H_y^\pm(x,y,z)$. From Eqs. (2.9) and (2.10), $E_x^\pm(x,y,z)$ and $H_y^\pm(x,y,z)$ are obtained from $A_x^\pm(x,y,z)$ as follows:

$$E_x^\pm(x,y,z) = ik\left(1 + \frac{1}{k^2}\frac{\partial^2}{\partial x^2}\right)A_x^\pm(x,y,z) \tag{69}$$

$$H_y^\pm(x,y,z) = \frac{\partial A_x^\pm(x,y,z)}{\partial z} \tag{70}$$

$A_x^\pm(x,y,z)$ and $E_x^\pm(x,y,z)$ are even functions of z, and $H_y^\pm(x,y,z)$ is an odd function of z. $H_y^\pm(x,y,z)$ is discontinuous across $z = 0$. This discontinuity causes a surface current density, as given by

$$\mathbf{J}(x,y,0) = \hat{z} \times \hat{y}[H_y^+(x,y,0) - H_y^-(x,y,0)] = -\hat{x}2H_y^+(x,y,0) \tag{71}$$

The current density is expressed as

$$\mathbf{J}(x,y,z) = -\hat{x}2H_y^+(x,y,0)\delta(z) \tag{72}$$

The complex power is obtained from Eq. (D18) as

$$P_C = -\frac{c}{2}\int_{-\infty}^\infty\int_{-\infty}^\infty\int_{-\infty}^\infty dxdydz E_x^\pm(x,y,z)(-2)H_y^{+*}(x,y,0)\delta(z)$$

$$= c\int_{-\infty}^\infty\int_{-\infty}^\infty dxdy E_x^\pm(x,y,0)H_y^{+*}(x,y,0) \tag{73}$$

Using Eqs. (66), (69), and (70), P_C given by Eq. (73) is determined as

$$P_C = c\int_{-\infty}^\infty\int_{-\infty}^\infty dxdy(-i)\exp(-kb)\frac{N}{a^2}\exp\left[\frac{2}{3}\left(\frac{2\pi a}{a}\right)^3\right]$$

$$\times \int_{-\infty}^\infty\int_{-\infty}^\infty dp_x dp_y\exp\left[-i2\pi(p_x x + p_y y)\right]$$

$$\times \exp\left[-\frac{i}{3}\left(\frac{2\pi}{a}\right)^3 p_x^3 + i\left(\frac{2\pi}{a}\right)^3 a^2 p_x\right]\exp\left[-\frac{i}{3}\left(\frac{2\pi}{a}\right)^3 p_y^3 + i\left(\frac{2\pi}{a}\right)^3 a^2 p_y\right]$$

$$\times ik\left(1 - \frac{4\pi^2 p_x^2}{k^2}\right)\frac{1}{\zeta}\exp(\zeta b)(i)\exp(-kb)\frac{N}{a^2}\exp\left[\frac{2}{3}\left(\frac{2\pi a}{a}\right)^3\right]$$

$$\times \int_{-\infty}^\infty\int_{-\infty}^\infty d\bar{p}_x d\bar{p}_y\exp\left[i2\pi(\bar{p}_x x + \bar{p}_y y)\right]$$

$$\times \exp\left[\frac{i}{3}\left(\frac{2\pi}{a}\right)^3 \bar{p}_x^3 - i\left(\frac{2\pi}{a}\right)^3 a^2\bar{p}_x\right]\exp\left[\frac{i}{3}\left(\frac{2\pi}{a}\right)^3 \bar{p}_y^3 - i\left(\frac{2\pi}{a}\right)^3 a^2\bar{p}_y\right]$$

$$\times (-i)\exp(\bar{\zeta}^* b) \tag{74}$$

In Eq. (74), $\bar{\zeta}$ is the same as ζ, with p_x and p_y replaced by $\bar{p}_x$ and $\bar{p}_y$, respectively. The integrations with respect to x and y are performed first to obtain $\delta(p_x - \bar{p}_x)\delta(p_y - \bar{p}_y)$. Then, the integrations with respect to $\bar{p}_x$ and $\bar{p}_y$, are carried out. The result is

$$P_C = \frac{ckN^2}{a^4}\exp(-2kb)\exp\left[\frac{4}{3}\left(\frac{2\pi a}{a}\right)^3\right]$$

$$\times \int_{-\infty}^{\infty}\int_{-\infty}^{\infty} dp_x dp_y \left(1 - \frac{4\pi^2 p_x^2}{k^2}\right)\frac{1}{\zeta}\exp[b(\zeta + \zeta^*)] \tag{75}$$

The integration variables are changed as $2\pi p_x = p\cos\phi$ and $2\pi p_y = p\sin\phi$. Then, Eq. (75) becomes

$$P_C = \frac{cN^2\exp(-2kb)}{4\pi^2 a^4}\exp\left[\frac{4}{3}\left(\frac{2\pi a}{a}\right)^3\right]$$

$$\times \int_0^\infty dp\, p \int_0^{2\pi} d\phi\left(1 - \frac{p^2\cos^2\phi}{k^2}\right)\frac{\exp[kb(\xi + \xi^*)]}{(1 - p^2/k^2)^{1/2}} \tag{76}$$

where $\xi = (1 - p^2/k^2)^{1/2}$. In Eq. (30), for the corresponding paraxial beam, the time-averaged real power is obtained. For $P_{re} = 2$ W, the normalization constant N is determined and substituted into Eq. (76), with the result that

$$P_C = \frac{w_{0,eq}^2}{\pi}\exp(-2kb)\int_0^\infty dp\, p \int_0^{2\pi} d\phi\left(1 - \frac{p^2\cos^2\phi}{k^2}\right)\frac{\exp[kb(\xi + \xi^*)]}{\xi} \tag{77}$$

This result is identical to that given by Eq. (4.23) for the basic full Gaussian wave. The complex power of the basic full modified Airy wave is identical to that for the basic full Gaussian wave, provided that for the former an equivalent waist is defined in accordance with that given by Eq. (60). Each wave is normalized such that the time-averaged real power in the corresponding paraxial beam is 2 W.

The reactive power of the basic full modified Airy wave is infinite. This result is to be expected. Omission of the small finite dimensions of the source causes the reactive power to become infinite, whether the source is real and is situated in the real space [14] or whether the source is virtual and is situated in the complex space [15]. For the basic full Gaussian wave, the virtual source is a point source situated in the complex space at $x = 0$, $y = 0$, and $|z| = ib$, where b is the Rayleigh distance. For the basic full modified Airy wave, the virtual source is an infinite series of point sources with an increasing order starting from zero. The source is situated in the complex space at $x = 0$, $y = 0$, and $|z| = ib$, where b is the equivalent Rayleigh distance.

By symmetry, for both the waves, the time-averaged power flowing in the $+z$ direction is the same as that in the $-z$ direction. In the paraxial limit, this power is 1 W. In Fig. 4.1, for both waves, the time-averaged power is shown as a function of kw_0; this power increases monotonically, reaching the paraxial limit of 1 W as kw_0 is increased. Here, w_0 is the waist at the input for the Gaussian wave and is the equivalent waist for the modified Airy wave. The analytical expressions for the radiation intensity $\Phi(\theta, \phi)$ and the time-averaged power

$P_{MA}^{\pm}$ transported by the Airy wave in the $+z$ or $-z$ direction are given, respectively, by Eqs. (4.11) and (4.16). The radiation intensity pattern is shown in Fig. 4.2. Thus, the propagation characteristics of the basic full modified Airy wave are identical to those of the basic full Gaussian wave, provided that the waist w_0 of the Gaussian wave is replaced by the equivalent waist for the basic full modified Airy wave.

5 Remarks

In conclusion, the fundamental Airy beam is introduced; this paraxial beam is not physically realizable. Therefore, the paraxial approximation leading to the fundamental Airy beam is not valid. The full-wave generalization of the fundamental Airy beam is developed to obtain the fundamental Airy wave. This physically realizable full wave has a radiation intensity distribution that is the same as that of a point electric dipole. The finite-energy fundamental Airy beam is presented. This paraxial beam is physically realizable and has a wavenumber distribution similar to that of a Gaussian beam. The complex space source theory of the finite-energy fundamental Airy beam is presented. When an appropriate equivalent waist is introduced, the propagation characteristics of the basic full modified Airy wave are identical to those of the basic full Gaussian wave. The source for the basic full Gaussian wave is a point electric dipole in the complex space. The source for the basic full modified Airy wave is an infinite series of higher-order point sources with an increasing order starting from zero. Thus, the basic full modified Airy wave carries the features of a dressed point electric dipole.

References

1. E. G. Kalnins and W. Miller Jr., "Lie theory and separation of variables. 5. The equations $iU_t + U_{xx} = 0$ and $iU_t + U_{xx} - c/x^2 U = 0$," *J. Math. Phys.* **15**, 1728–1737 (1974).

2. M. V. Berry and N. L. Balazs, "Nonspreading wave packets," *Am. J. Phys.* **47**, 264–267 (1979).

3. I. M. Besieris, A. M. Shaarawi, and R. W. Ziolkowski, "Nondispersive accelerating wave packets," *Am. J. Phys.* **62**, 519–521 (1994).

4. K. Unnikrishnan and A. R. P. Rau, "Uniqueness of the Airy packet in quantum mechanics," *Am. J. Phys.* **64**, 1034–1036 (1996).

5. G. A. Siviloglou and D. N. Christodoulides, "Accelerating finite energy Airy beams," *Opt. Lett.* **32**, 979–981 (2007).

6. I. M. Besieris and A. M. Shaarawi, "A note on an accelerating finite energy Airy beams," *Opt. Lett.* **32**, 2447–2449 (2007).

7. G. A. Siviloglou, J. Broky, A. Dogariu, and D. N. Christoloulides, "Observation of accelerating Airy beams," *Phys. Rev. Lett.* **99**, 213901 (2007).

8. M. A. Bandres and J. C. Gutierrez-Vega, "Airy-Gauss beams and their transformation by paraxial optical systems," *Opt. Express,* **15**, 16719–16728 (2007).

9. S. Yan, B. Yao, M. Lei, D. Dan, Y. Yang, and P. Gao, "Virtual source for an Airy beam," *Opt. Lett.* **37**, 4774–4776 (2012).

10. G. A. Deschamps, "Gaussian beam as a bundle of complex rays," *Electron. Lett.* **7**, 684–685 (1971).

11. M. A. Bandres and J. C. Gutierrez-Vega, "Higher-order complex source for elegant Laguerre-Gaussian waves," *Opt. Lett.* **29**, 2213–2215 (2004).

12. Y. Zhang, Y. Song, Z. Chen, J. Ji, and Z. Shi, "Virtual sources for cosh-Gaussian beam," *Opt. Lett.* **32**, 292–294 (2007).

13. S. R. Seshadri, "Full-wave generalizations of the fundamental Gaussian beam," *J. Opt. Soc. Am. A* **26**, 2515–2520 (2009).

14. S. R. Seshadri, "Constituents of power of an electric dipole of finite size," *J. Opt. Soc. Am. A* **25**, 805–810 (2008).

15. S. R. Seshadri, "Reactive power in the full Gaussian light wave," *J. Opt. Soc. Am. A* **26**, 2427–2433 (2009).

Green's function for the Helmholtz equation

A.1 Three-dimensional scalar Green's function

The three-dimensional free-space Green's function, $G(x, y, z)$, is defined by the inhomogeneous Helmholtz equation

$$\left(\frac{\partial^2}{\partial x^2} + \frac{\partial^2}{\partial y^2} + \frac{\partial^2}{\partial z^2} + k^2\right) G(x, y, z) = -\delta(x)\delta(y)\delta(z) \tag{A1}$$

where k is the wavenumber and $\delta(\ )$ is the delta function. The following three-dimensional Fourier transform pair is used:

$$G(x, y, z) = \int_{-\infty}^{\infty} \int_{-\infty}^{\infty} \int_{-\infty}^{\infty} \overline{G}(p_x, p_y, p_z) \exp(-i2\pi \, \mathbf{p} \cdot \mathbf{r}) dp_x dp_y dp_z \tag{A2}$$

and

$$\overline{G}(p_x, p_y, p_z) = \int_{-\infty}^{\infty} \int_{-\infty}^{\infty} \int_{-\infty}^{\infty} G(x, y, z) \exp(i2\pi \mathbf{p} \cdot \mathbf{r}) dx \, dy dz \tag{A3}$$

where

$$\mathbf{p} = \hat{x} p_x + \hat{y} p_y + \hat{z} p_z \tag{A4}$$

and

$$\mathbf{r} = \hat{x} x + \hat{y} y + \hat{z} z \tag{A5}$$

In view of Eqs. (A2) and (A3), the delta function has the representation

$$\delta(x) = \int_{-\infty}^{\infty} \exp(-i2\pi \, p_x x) dp_x \tag{A6}$$

and similarly for $\delta(y)$ and $\delta(z)$ as well. Substituting Eq. (A2) and the equations for the delta functions similar to Eq. (A6) into Eq. (A1) yields

$$[-4\pi^2(p_x^2 + p_y^2 + p_z^2) + k^2]\overline{G}(p_x, p_y, p_z) = -1 \tag{A7}$$

When $\overline{G}(p_x, p_y, p_z)$ from Eq. (A7) is substituted into Eq. (A2), it is found that

$$G(x, y, z) = \frac{1}{4\pi^2} \int_{-\infty}^{\infty} \int_{-\infty}^{\infty} \int_{-\infty}^{\infty} \frac{\exp(-i2\pi\, \mathbf{p} \cdot \mathbf{r})}{(p_x^2 + p_y^2 + p_z^2 - k^2/4\pi^2)}\, dp_x dp_y dp_z \tag{A8}$$

Another rectangular coordinate system (X, Y, Z) having the same origin as the rectangular coordinate system (x, y, z) is introduced. The Z axis is chosen to be in the direction of the radius vector $\mathbf{r}$, that is $\mathbf{r} = \hat{Z}r$. Then, $\mathbf{p}$ in Eq. (A4) is expressed as

$$\mathbf{p} = \hat{X}p_X + \hat{Y}p_Y + \hat{Z}p_Z \tag{A9}$$

Spherical coordinates with respect to the Z axis of the (X, Y, Z) coordinate system are used to represent p_X, p_Y, and p_Z as follows:

$$p_X = p\sin\Theta\cos\Phi, \quad p_Y = p\sin\Theta\sin\Phi, \quad p_Z = p\cos\Theta \tag{A10}$$

where

$$p = |\mathbf{p}| = (p_X^2 + p_Y^2 + p_Z^2)^{1/2} \tag{A11}$$

Equations (A4) and (A10) are substituted into Eq. (A8). It is noted that $\mathbf{p} \cdot \mathbf{r} = rp\cos\Theta$. Then, Eq. (A8) can be recast as

$$G(x, y, z) = \frac{1}{4\pi^2} \int_0^{\infty} dp\, \frac{p^2}{(p^2 - k^2/4\pi^2)}$$
$$\times \int_0^{\pi} d\Theta\, \sin\Theta\exp(-i2\pi\, rp\cos\Theta) \int_0^{2\pi} d\Phi \tag{A12}$$

The integrations with respect to Φ and Θ are performed. The result is

$$G(x, y, z) = \frac{1}{4\pi^2 ir} \int_0^{\infty} dp\, \frac{p[\exp(i2\pi\, rp) - \exp(-i2\pi\, rp)]}{(p^2 - k^2/4\pi^2)} \tag{A13}$$

For $\exp(-i2\pi rp)$, the integration variable is changed from p to $-p$. Then, Eq. (A13) changes into the form

$$G(x, y, z) = \frac{1}{4\pi^2 ir} \int_{-\infty}^{\infty} dp\, \frac{p\exp(i2\pi\, rp)}{(p^2 - k^2/4\pi^2)} \tag{A14}$$

If p is real, the integral is not defined since the integrand becomes infinite at $p = \pm k/2\pi$. To overcome this difficulty, the medium is assumed to have a small loss. The fields are assumed to have a harmonic time dependence of the form $\exp(-i\omega t)$, where $\omega/2\pi$ is the wave frequency. For the assumed time dependence, when there is loss, k becomes complex, with a small positive imaginary part. The domain of p is extended to complex values, $p = p_r + ip_i$, where p_r and p_i are the real and imaginary parts of p (Fig. 1). The singularity at $k/2\pi$ moves up $(p_i > 0)$ and that at $-k/2\pi$ moves down $(p_i < 0)$. The contour of integration is along the real p axis. The integral in Eq. (A14) is defined well. As the losses are reduced, the singularity at $k/2\pi$ moves down to the real axis, and in order not to change the value of the integral, the contour of integration is analytically continued by way of the lower half of the complex p plane. Similarly, in the limit of losses going to zero, the singularity at $-k/2\pi$ moves up to the real axis and the value of the integral remains unchanged if the contour of integration is analytically continued by way of the upper half of the complex p plane. Since $r > 0$, if $p_i > 0$, the exponent in Eq. (A14) has a negative real part. Thus, along the infinite semicircle in the upper half of the complex p plane, the integral has no contribution. Therefore, the contour of integration is closed by the infinite semicircle in the upper half of the complex p plane without changing the value of the integral. For $r > 0$, the contribution to the integral in Eq. (A14) arises only from the simple pole at $p = k/2\pi$. The contour is along the real axis, continued by way of the upper half of the p plane at $p = -k/2\pi$ and by way of the lower half of the p plane at $p = k/2\pi$, and closed by the infinite semicircle in the upper half of the p plane. The contour of integration encircles the pole at $p = k/2\pi$ in the counterclockwise direction. Therefore, the value of the integral in Eq. (A14) equals $+2\pi i$ times the residue at the simple pole at $p = k/2\pi$. The residue is obtained by multiplying the integrand by $(p - k/2\pi)$ and setting $p = k/2\pi$ in the result. Hence, it is found from Eq. (A14) that

$$G(x,y,z) = \frac{1}{4\pi^2 i\, r} .2\pi\, i\, . \frac{\exp(ikr)}{2} = \frac{\exp(ikr)}{4\pi r} \tag{A15}$$

It follows from Eq. (A5) that the distance r from the origin to the point of observation is given by

$$r = \left(x^2 + y^2 + z^2\right)^{1/2} \tag{A16}$$

A.2 Fourier transform of scalar Green's function

The two-dimensional Fourier transform of the free-space Green's function, $G(x,y,z)$, is deduced directly from Eq. (A1). The two-dimensional Fourier transform pair is defined as follows:

$$G(x,y,z) = \int_{-\infty}^{\infty} \int_{-\infty}^{\infty} \overline{G}(p_x, p_y, z) \exp\left[-i2\pi\,(p_x x + p_y y)\right] dp_x dp_y \tag{A17}$$

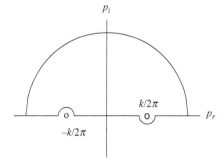

Fig. 1. Contour of integration in the complex p plane.

and

$$\overline{G}(p_x, p_y, z) = \int_{-\infty}^{\infty} \int_{-\infty}^{\infty} G(x, y, z) \exp\left[i2\pi \left(p_x x + p_y y\right)\right] dx \, dy \tag{A18}$$

When Eq. (A17) and the representations for the delta functions, as given by Eq. (A6), are substituted into Eq. (A1), it is found that

$$\left(\frac{\partial^2}{\partial z^2} + \zeta^2\right)\overline{G}(p_x, p_y, z) = -\delta(z) \tag{A19}$$

where

$$\zeta = [k^2 - 4\pi^2(p_x^2 + p_y^2)]^{1/2} \tag{A20}$$

For $z \neq 0$, the solutions of Eq. (A19) are expressed as

$$\overline{G}(p_x, p_y, z) = A^+ \exp(i\zeta z) \quad \text{for } z > 0 \tag{A21}$$

$$= A^- \exp(-i\zeta z) \quad \text{for } z < 0 \tag{A22}$$

$\overline{G}(p_x, p_y, z)$ is continuous at $z = 0$ and

$$\frac{\partial}{\partial z}\overline{G}(p_x, p_y, 0+) - \frac{\partial}{\partial z}\overline{G}(p_x, p_y, 0-) = -1 \tag{A23}$$

Therefore,

$$A^+ = A^- = i/2\zeta \tag{A24}$$

From Eqs. (A21), (A22), and (A24), $\overline{G}(p_x, p_y, z)$ is obtained as

$$\overline{G}(p_x, p_y, z) = \frac{i}{2\zeta} \exp(i\zeta|z|) \tag{A25}$$

From Eqs. (A17) and (A25), the three-dimensional free-space Green's function is found to have the following two-dimensional Fourier transform representation:

$$G(x, y, z) = \int_{-\infty}^{\infty} \int_{-\infty}^{\infty} \exp\left[-i2\pi \left(p_x x + p_y y\right)\right] \frac{i}{2\zeta} \exp(i\zeta|z|) dp_x dp_y \tag{A26}$$

A.3 Bessel transform of scalar Green's function

For the cylindrical coordinate system (ρ, ϕ, z), x and y are related to ρ and ϕ as follows:

$$x = \rho \cos\phi, \quad y = \rho \sin\phi \tag{A27}$$

Equation (A1) transforms as

$$\left(\frac{\partial^2}{\partial \rho^2} + \frac{1}{\rho} \frac{\partial}{\partial \rho} + \frac{1}{\rho^2} \frac{\partial^2}{\partial \phi^2} + \frac{\partial^2}{\partial z^2} + k^2 \right) G(\rho, z) = -\frac{\delta(\rho)}{2\pi\rho} \delta(z) \tag{A28}$$

where $\rho = (x^2 + y^2)^{1/2}$. When the right-hand side of Eq. (A1) is multiplied by the element of volume $dxdydz$ and integrated with respect to x, y, and z from $-\infty$ to ∞, the result is -1. Similarly, when the right-hand side of Eq. (A28) is multiplied by the element of volume $d\rho\rho d\phi dz$ in the cylindrical coordinates and integrated with respect to z from $-\infty$ to ∞, with respect to ϕ from 0 to 2π and with respect to ρ from 0 to ∞, the result is again -1. Thus, the strengths of the sources given by Eqs. (A1) and (A28) are the same. The source term in Eq. (A28) is independent of ϕ; therefore, $G(\rho, z)$ is independent of ϕ, as indicated. The Bessel transform pair is defined as follows:

$$G(\rho, z) = \int_0^\infty \overline{G}(\eta, z) J_0(\eta\rho) \eta d\eta \tag{A29}$$

and

$$\overline{G}(\eta, z) = \int_0^\infty G(\rho, z) J_0(\eta\rho) \rho d\rho \tag{A30}$$

where $J_0(\ )$ is the Bessel functions of order 0. In view of Eqs. (A29) and (A30), $\delta(\rho)/\rho$ has the following representation:

$$\frac{\delta(\rho)}{\rho} = \int_0^\infty J_0(\eta\rho) \eta d\eta \tag{A31}$$

Since $G(\rho, z)$ is independent of ϕ, the $(1/\rho^2)(\partial^2/\partial\phi^2)$ term is omitted. Substituting Eqs. (A29) and (A31) into Eq. (A28) and noting that

$$\left(\frac{\partial^2}{\partial \rho^2} + \frac{1}{\rho} \frac{\partial}{\partial \rho} + \eta^2 \right) J_0(\eta\rho) = 0 \tag{A32}$$

yields

$$\left(\frac{\partial^2}{\partial z^2} + \zeta^2 \right) \overline{G}(\eta, z) = -\frac{\delta(z)}{2\pi} \tag{A33}$$

where

$$\zeta = (k^2 - \eta^2)^{1/2} \tag{A34}$$

Equation (A33) is solved in the same way as Eq. (A19), with the result that

$$\overline{G}(\eta, z) = \frac{i \exp(i\zeta|z|)}{4\pi\zeta} \tag{A35}$$

Using Eqs. (A29) and (A35), the Bessel transform representation of $G(\rho, z)$ is obtained as

$$G(\rho, z) = \frac{i}{4\pi} \int_0^\infty d\eta \eta J_0(\eta\rho) \frac{\exp(i\zeta|z|)}{\zeta} \tag{A36}$$

Equation (A36) is known as Sommerfeld's formula [1].

Reference

1. W. Magnus and F. Oberhettinger, *Functions of Mathematical Physics* (Chelsea Publishing Company, New York, 1954), p. 34.

An integral

Consider the integral

$$I(a, b) = \int_{-\infty}^{\infty} dx \exp\left(iax - bx^2\right) \tag{B1}$$

where a is real and the real part of b is positive. Then, $I(a, b)$ is finite. The argument of the exponential function is rearranged as

$$-bx^2 + iax = -\frac{a^2}{4b} - b\left(x - \frac{ia}{2b}\right)^2 \tag{B2}$$

The integration variable is changed as

$$y = \sqrt{b}\left(x - \frac{ia}{2b}\right) \tag{B3}$$

By using Eqs. (B2) and (B3), Eq. (B1) is transformed as

$$I(a, b) = \frac{\exp\left(-a^2/4b\right)}{\sqrt{b}} \int_{L}^{U} dy \exp\left(-y^2\right) \tag{B4}$$

where $L = -\infty - ia/2\sqrt{b}$ and $U = \infty - ia/2\sqrt{b}$. The contour of integration can be shifted up to coincide with the real axis without changing the value of the integral. The limits of the integral become the same as in Eq. (B1). Since

$$\int_{-\infty}^{\infty} dy \exp\left(-y^2\right) = \sqrt{\pi} \tag{B5}$$

it is found from Eq. (B4) that

$$I(a, b) = \sqrt{\frac{\pi}{b}}\exp\left(-\frac{a^2}{4b}\right) \tag{B6}$$

Green's function for the paraxial equation

C.1 Paraxial approximation

External to the source region, the three-dimensional free-space Green's function satisfies the following Helmholtz equation:

$$\left(\frac{\partial^2}{\partial x^2} + \frac{\partial^2}{\partial y^2} + \frac{\partial^2}{\partial z^2} + k^2\right) G(x,y,z) = 0 \tag{C1}$$

For a nearly plane wave having a small range of wave vectors about the propagation direction, the plane-wave phase factor is separated out as

$$G_p(x,y,z) = \exp(\pm ikz) g_p(x,y,z) \tag{C2}$$

where $\pm$ denotes the $\pm z$ direction of propagation. The subscript p is used to indicate paraxial. For a plane wave, $g_p(x,y,z)$ is a constant. When the wave is a nearly plane wave or a beam, $g_p(x,y,z)$ is a slowly varying function of its arguments. Substituting Eq. (C2) into Eq. (C1) leads to the following differential equation satisfied by $g_p(x,y,z)$:

$$\left(\frac{\partial^2}{\partial x^2} + \frac{\partial^2}{\partial y^2} \pm 2ik\frac{\partial}{\partial z} + \frac{\partial^2}{\partial z^2}\right) g_p(x,y,z) = 0 \tag{C3}$$

Let w_0 and b be the scale lengths of variation of $g_p(x,y,z)$ in the transverse (x,y) directions and the longitudinal direction, respectively, where w_0 is the beam waist at the input $(z=0)$ plane and $b = \frac{1}{2}kw_0^2$ is the basic Rayleigh distance. The use of the normalized variables [1]

$$x_n = x/w_0, \quad y_n = y/w_0 \quad \text{and} \quad z_n = z/b \tag{C4}$$

changes Eq. (C3) as

$$\left(\frac{\partial^2}{\partial x_n^2} + \frac{\partial^2}{\partial y_n^2} \pm 4i\frac{\partial}{\partial z_n} + \frac{4}{k^2 w_0^2}\frac{\partial^2}{\partial z_n^2}\right) g_p(x_n,y_n,z_n) = 0 \tag{C5}$$

Usually, w_0 is very large compared to the wavelength $2\pi/k$; therefore, $kw_0 \gg 1$. The paraxial approximation corresponds to $kw_0 \gg 1$ or $f_0 \ll 1$, where

$$f_0 = 1/kw_0 \tag{C6}$$

In the paraxial approximation, the $\partial^2/\partial z^2$ term can be omitted, and from Eq. (C3), the paraxial wave equation is obtained as

$$\left(\frac{\partial^2}{\partial x^2} + \frac{\partial^2}{\partial y^2} \pm 2ik\frac{\partial}{\partial z} \right) g_p(x,y,z) = 0 \tag{C7}$$

C.2 Green's function $G_p(x, y, z)$

For the paraxial equation, the input value $g_p(x,y,0)$ is specified and the value $g_p(x,y,z)$ for $|z| > 0$ is determined. The input value of $g_p(x,y,z)$ is specified as

$$g_p(x,y,0) = \delta(x)\delta(y) \tag{C8}$$

The two-dimensional Fourier transform of Eq. (C8) is found from Eq. (A18) as

$$\bar{g}_p(p_x,p_y,0) = 1 \tag{C9}$$

Substituting the two-dimensional Fourier transform representation of $g_p(x,y,z)$, as given by Eq. (A17), into Eq. (C7) leads to the differential equation for $\bar{g}_p(p_x,p_y,z)$ as

$$\left[-4\pi^2(p_x^2 + p_y^2) \pm 2ik\frac{\partial}{\partial z} \right] \bar{g}_p(p_x,p_y,z) = 0 \tag{C10}$$

The solution of Eq. (C10) together with Eq. (C9) is

$$\bar{g}_p(p_x,p_y,z) = \exp\left[-\pi^2 w_0^2(p_x^2 + p_y^2)\frac{i|z|}{b} \right] \tag{C11}$$

When the inverse Fourier transform of Eq. (C11) is determined using Eq. (A17) and the resulting integrals are evaluated using Eqs. (B1) and (B6), the result is

$$g_p(x,y,z) = -\frac{ik}{2\pi|z|}\exp\left[\frac{ik(x^2+y^2)}{2|z|} \right] \tag{C12}$$

From Eqs. (C2) and (C12), the Green's function for the paraxial equation is obtained as

$$G_p(x,y,z) = \exp(\pm ikz)\left(-\frac{ik}{2\pi|z|} \right)\exp\left[\frac{ik(x^2+y^2)}{2|z|} \right] \tag{C13}$$

The slowly varying amplitude part of $G_p(x,y,z)$ is given by $g_p(x,y,z)$.

The paraxial approximation of $G(x,y,z)$ can be found directly from Eqs. (A15) and (A16). The paraxial region corresponds to $(x^2 + y^2)/z^2 \ll 1$. When r in Eq. (A16) is expanded into a power series in $(x^2 + y^2)/z^2$, the first two terms are obtained as

$$r = |z| + (x^2 + y^2)/2|z| \tag{C14}$$

In Eq. (A15), if r in the amplitude is replaced by the first term in Eq. (C14) and r in the phase is replaced by the first two terms in Eq. (C14), the direct paraxial approximation of $G(x,y,z)$ given by Eq. (A15) is found as

$$G(x,y,z) = \exp(\pm ikz)\left(\frac{1}{4\pi|z|}\right)\exp\left[\frac{ik(x^2 + y^2)}{2|z|}\right] \tag{C15}$$

The paraxial approximation of $G(x,y,z)$ obtained directly differs by the factor $1/(-2ik)$ from the Green's function derived by the use of the paraxial equation.

C.3 Fourier transform $\overline{G}_p(p_x, p_y, z)$

From Eqs. (C2) and (C11), the Fourier transform of the Green's function for the paraxial equation is obtained as

$$\overline{G}_p(p_x, p_y, z) = \exp(\pm ikz)\exp\left[-\pi^2 w_0^2(p_x^2 + p_y^2)\frac{i|z|}{b}\right] \tag{C16}$$

In Eq. (A20), the paraxial approximation corresponds to $4\pi^2(p_x^2 + p_y^2)/k^2 \ll 1$. When ζ in Eq. (A20) is expanded into a power series in $4\pi^2(p_x^2 + p_y^2)/k^2$, the first two terms are found as

$$\zeta = k - \pi^2 w_0^2(p_x^2 + p_y^2)/b \tag{C17}$$

In Eq. (A25), if ζ in the amplitude is replaced by the first term in Eq. (C17) and ζ in the phase is replaced by the first two terms in Eq. (C17), the direct paraxial approximation of $\overline{G}(p_x, p_y, z)$ in Eq. (A25) is given by

$$\overline{G}(p_x, p_y, z) = \exp(\pm ikz)\left(\frac{1}{-2ik}\right)\exp\left[-\pi^2 w_0^2(p_x^2 + p_y^2)\frac{i|z|}{b}\right] \tag{C18}$$

As is to be expected, the paraxial approximation of $\overline{G}(p_x, p_y, z)$ derived directly differs by the factor $1/(-2ik)$ from $\overline{G}_p(p_x, p_y, z)$ determined by the use of the paraxial equation.

C.4 Bessel transform $\overline{G}_p(\eta, z)$

A similar relation exists for the Bessel transforms of the two Green's functions. Cylindrical coordinates are used, and the input value of the slowly varying amplitude $g_p(\rho, z)$ is specified as

$$g_p(\rho, 0) = \delta(\rho)/2\pi\rho \tag{C19}$$

By using Eq. (A30), the Bessel transform of Eq. (C19) is found as

$$\overline{g}_p(\eta, 0) = 1/2\pi \tag{C20}$$

The input value of $g_p(\rho, z)$ is independent of ϕ; therefore, $g_p(\rho, z)$ is independent of ϕ for all z. The cylindrically symmetric form of the paraxial equation as given by Eq. (C7) is transformed as

$$\left(\frac{\partial^2}{\partial \rho^2} + \frac{1}{\rho} \frac{\partial}{\partial \rho} \pm 2ik \frac{\partial}{\partial z} \right) g_p(\rho, z) = 0 \tag{C21}$$

When $g_p(\rho, z)$ is expressed as the inverse Bessel transform of $\overline{g}_p(\eta, z)$ as given by Eq. (A29), and Eq. (A32) is used, the following differential equation for $\overline{g}_p(\eta, z)$ is obtained from Eq. (C21):

$$\left(-\eta^2 \pm 2ik \frac{\partial}{\partial z} \right) \overline{g}_p(\eta, z) = 0 \tag{C22}$$

The use of Eq. (C20) enables the solution of Eq. (C22) to be expressed as

$$\overline{g}_p(\eta, z) = \frac{1}{2\pi} \exp\left(-\frac{\eta^2 w_0^2}{4} \frac{i|z|}{b} \right) \tag{C23}$$

Then, $g_p(\rho, z)$ is found as

$$g_p(\rho, z) = \frac{1}{2\pi} \int_0^\infty d\eta \eta J_0(\eta \rho) \exp\left(-\frac{\eta^2 w_0^2}{4} \frac{i|z|}{b} \right) \tag{C24}$$

The integral is evaluated by the use of the formula {see p. 35 in [2]}

$$\int_0^\infty dt t J_0(at) \exp(-p^2 t^2) = \frac{1}{2p^2} \exp\left(-\frac{a^2}{4p^2} \right) \tag{C25}$$

with the result that

$$g_p(\rho, z) = -\frac{ik}{2\pi|z|} \exp\left(\frac{ik\rho^2}{2|z|} \right) \tag{C26}$$

From Eqs. (C2) and (C26), the Green's function for the paraxial equation is found as

$$G_p(\rho, z) = \exp(\pm ikz) \left(-\frac{ik}{2\pi|z|} \right) \exp\left(\frac{ik\rho^2}{2|z|} \right) \tag{C27}$$

From Eqs. (C2) and (C23), the Bessel transform of the Green's function of the paraxial equation is obtained as

$$\overline{G}_p(\eta, z) = \exp(\pm ikz) \frac{1}{2\pi} \exp\left(-\frac{\eta^2 w_0^2}{4} \frac{i|z|}{b} \right) \tag{C28}$$

In Eq. (A34), the paraxial approximation corresponds to $\eta^2/k^2 \ll 1$. When ζ in Eq. (A34) is expanded into a power series in η^2/k^2, the first two terms are found as

$$\zeta = k - \eta^2/2k \tag{C29}$$

In Eq. (A35), if ζ in the amplitude is replaced by the first term in Eq. (C29) and ζ in the phase is replaced by the first two terms in Eq. (C29), the direct paraxial approximation of $\overline{G}(\eta, z)$ in Eq. (A35) is determined as

$$\overline{G}(\eta, z) = \exp(\pm ikz)\left(\frac{1}{-4\pi ik}\right)\exp\left(-\frac{\eta^2 w_0^2}{4}\frac{i|z|}{b}\right) \tag{C30}$$

A comparison of Eq. (C30) with Eq. (C28) shows that the Bessel transform $\overline{G}(\eta, z)$ of the paraxial approximation of the Green's function derived directly differs by the factor $1/(-2ik)$ from the Bessel transform $\overline{G}_p(\eta, z)$ of the paraxial approximation of the Green's function determined by the use of the paraxial equation.

Thus, there is a difference in the definitions of the Green's functions for the Helmholtz equation and the paraxial equation. This difference is removed by the introduction of an appropriate excitation constant for the solution obtained from the Helmholtz equation. Then, the paraxial approximation of the exact solution found from the Helmholtz equation is the same as the paraxial approximation deduced from the paraxial equation.

References

1. M. Lax, W. H. Louisell, and W. B. McKnight, "From Maxwell to paraxial wave optics," *Phys. Rev. A* **11**, 1365–1370 (1975).

2. W. Magnus and F. Oberhettinger, *Functions of Mathematical Physics* (Chelsea Publishing Company, New York, 1954), p. 35.

APPENDIX D

Electromagnetic fields

D.1 Poynting vector and generated power

In free space (μ_0, ε_0), the electric field, the magnetic field, the magnetic current density, the electric current density, the equivalent magnetic charge density, and the electric charge density are denoted by $\tilde{\mathbf{E}}$, $\tilde{\mathbf{H}}$, $\tilde{\mathbf{J}}_m$, $\tilde{\mathbf{J}}_e$, $\tilde{\rho}_m$, and $\tilde{\rho}_e$, respectively. The normalized fields and the source quantities as defined by

$$\mathbf{E} = (\varepsilon_0)^{1/2}\tilde{\mathbf{E}}, \quad \mathbf{H} = (\mu_0)^{1/2}\,\tilde{\mathbf{H}}, \quad \mathbf{J}_m = (\varepsilon_0)^{1/2}\tilde{\mathbf{J}}_m$$

$$\mathbf{J}_e = (\mu_0)^{1/2}\tilde{\mathbf{J}}_e, \quad \rho_m = (\mu_0)^{-1/2}\tilde{\rho}_m, \quad \rho_e = (\varepsilon_0)^{-1/2}\tilde{\rho}_e$$

satisfy the time-dependent Maxwell's equations:

$$\nabla \times \mathbf{E} = -\frac{1}{c}\frac{\partial \mathbf{H}}{\partial t} - \mathbf{J}_m \tag{D1}$$

$$\nabla \times \mathbf{H} = \frac{1}{c}\frac{\partial \mathbf{E}}{\partial t} + \mathbf{J}_e \tag{D2}$$

$$\nabla \cdot \mathbf{H} = \rho_m \tag{D3}$$

$$\nabla \cdot \mathbf{E} = \rho_e \tag{D4}$$

where c is the free-space electromagnetic wave velocity {see pp. 207–227 in [1]}. The arguments of all the functions in Eqs. (D1)–(D4) are the position vector $\mathbf{r}$ and the time t. From Eqs. (D1) and (D2), the following relation is deduced:

$$\nabla \cdot c(\mathbf{E} \times \mathbf{H}) = -\frac{\partial}{\partial t}\left(\frac{1}{2}\mathbf{E}\cdot\mathbf{E} + \frac{1}{2}\mathbf{H}\cdot\mathbf{H}\right) - c(\mathbf{E}\cdot\mathbf{J}_e + \mathbf{H}\cdot\mathbf{J}_m) \tag{D5}$$

By integrating Eq. (D5) throughout a volume V bounded by the surface A, the conservation relation is obtained as

$$\oint_A dA \cdot c(\mathbf{E} \times \mathbf{H}) = -\frac{\partial}{\partial t} \int_V dV \left(\frac{1}{2} \mathbf{E} \cdot \mathbf{E} + \frac{1}{2} \mathbf{H} \cdot \mathbf{H} \right) - c \int_{V_s} dV (\mathbf{E} \cdot \mathbf{J}_e + \mathbf{H} \cdot \mathbf{J}_m) \qquad (D6)$$

where the volume V_s is a part of V and contains the electric and the magnetic current sources. The first term on the right-hand side of Eq. (D6) is the rate of decrease of the electromagnetic energy inside the volume V, and the second term is the electromagnetic power generated inside the volume V_s. The energy conservation law shows that $c(\mathbf{E} \times \mathbf{H})$ is the Poynting vector that is the outward power flow per unit area across the surface A.

The fields are now assumed to have a harmonic time dependence of the form $\exp(-i\omega t)$, where $\omega/2\pi$ is the wave frequency. The form of the field is

$$f(\mathbf{r}, t) = f(\mathbf{r}) \exp(-i\omega t) \qquad (D7)$$

where $f(\mathbf{r})$ is the complex phasor quantity, the real part of the right-hand side is implied, and $\exp(-i\omega t)$ is suppressed. Then, the time-harmonic forms of Eqs. (D1) and (D2) are given by

$$\nabla \times \mathbf{E}(\mathbf{r}) = i\frac{\omega}{c}\mathbf{H}(\mathbf{r}) - \mathbf{J}_m(\mathbf{r}) \qquad (D8)$$

$$\nabla \times \mathbf{H}(\mathbf{r}) = -i\frac{\omega}{c}\mathbf{E}(\mathbf{r}) + \mathbf{J}_e(\mathbf{r}) \qquad (D9)$$

The time average of the Poynting vector $\mathbf{S}(\mathbf{r}, t)$ and the power density generated $p_g(\mathbf{r}, t)$ are found as

$$\langle \mathbf{S}(\mathbf{r}, t) \rangle = \mathbf{S}(\mathbf{r}) = \frac{1}{2} \text{Re}[c\mathbf{E}(\mathbf{r}) \times \mathbf{H}^*(\mathbf{r})] \qquad (D10)$$

and

$$\langle p_g(\mathbf{r}, t) \rangle = p_g(\mathbf{r}) = -\frac{1}{2} \text{Re}\, c[\mathbf{E}(\mathbf{r}) \cdot \mathbf{J}_e^*(\mathbf{r}) + \mathbf{H}(\mathbf{r}) \cdot \mathbf{J}_m^*(\mathbf{r})] \qquad (D11)$$

where Re stands for the real part and the asterisk denotes complex conjugation. In the same way that Eq. (D5) was derived from Eqs. (D1) and (D2), it is deduced from Eqs. (D8) and (D9) that

$$\nabla \cdot \mathbf{S}(\mathbf{r}) = p_g(\mathbf{r}) \qquad (D12)$$

where $\mathbf{S}(\mathbf{r})$ and $p_g(\mathbf{r})$ are substituted from Eqs. (D10) and (D11), respectively. By integrating Eq. (D12) throughout a volume V enclosed by the surface A, we find that

$$\oint_A \mathbf{S}(\mathbf{r}) \cdot d\mathbf{A} = \int_{V_s} p_g(\mathbf{r}) dV \qquad (D13)$$

The time-averaged value of the power generated inside the volume V_s is equal to the time-averaged value of the power flowing out of a large surface A enclosing the volume V_s. Thus, the time-averaged value of the power created by the current sources can be determined by calculations made far from the sources; the time-averaged Poynting vector is obtained at large distances from the sources, from which the time-averaged power flowing out of a large surface enclosing the current sources is found; and this result, as stated in Eq. (D13), is equal to the time-averaged value of the power created by the current sources.

The instantaneous power generated cannot be obtained by calculations made far from the current distributions; it must be determined by integration over the volume V_s containing the current sources. For simple current sources, the instantaneous power can be expressed in a form that is amenable to a physical interpretation [2,3]. With a view to simplifying the details, the instantaneous powers $P_{ge}(t)$ and $P_{gm}(t)$ generated by the electric and the magnetic current sources, respectively, are considered separately. For the frequency $\omega/2\pi$, $\mathbf{E}(\mathbf{r}, t)$ and $\mathbf{J}_e(\mathbf{r}, t)$ are expressed in detail as

$$\mathbf{E}(\mathbf{r}, t) = \frac{1}{2}[\mathbf{E}(\mathbf{r}, \omega) \exp(-i\omega t) + \mathbf{E}^*(\mathbf{r}, \omega) \exp(i\omega t)] \tag{D14}$$

and

$$\mathbf{J}_e(\mathbf{r}, t) = \frac{1}{2}[\mathbf{J}_e(\mathbf{r}, \omega) \exp(-i\omega t) + \mathbf{J}_e^*(\mathbf{r}, \omega) \exp(i\omega t)] \tag{D15}$$

For simplicity, the electric current density is assumed to be of the form

$$\mathbf{J}_e(\mathbf{r}, \omega) = \mathbf{u}_e(\mathbf{r}, \omega) J_e(\mathbf{r}, \omega) \exp(i\theta_e) \tag{D16}$$

where $\mathbf{u}_e(\mathbf{r}, \omega)$ is a real unit vector and $J_e(\mathbf{r}, \omega)$ and θ_e are real numbers. In addition, the electric current source is assumed to be localized, with θ_e being independent of $\mathbf{r}$. Using Eqs. (D14)–(D16), $P_{ge}(t)$ is expressed as

$$P_{ge}(t) = P_{re}(\omega) 2\cos^2(\omega t - \theta_e) + P_{ie}(\omega)\sin[2(\omega t - \theta_e)] \tag{D17}$$

where

$$P_{Ce}(\omega) = P_{re}(\omega) + iP_{ie}(\omega) = -\frac{c}{2}\int_{V_s} \mathbf{E}(\mathbf{r}, \omega) \cdot \mathbf{J}_e^*(\mathbf{r}, \omega) dV \tag{D18}$$

For a localized electric current source, the first term in Eq. (D17) represents the radiative power irretrievably propagated and absorbed by a sphere at infinity, and the second term represents the reactive power that oscillates between the source and the field. The time average of the irretrievable or real power is equal to $P_{re}(\omega)$, and the time average of the reactive power is zero. At every instant, the total instantaneous power is equal to the algebraic sum of the two terms in Eq. (D17). The physical basis for this separation is similar to that in circuit theory {see pp. 67–71 in [1]}. There, the first term would represent the instantaneous power delivered to the resistor, and the second term would represent the power associated with the discharging and charging of a capacitor. $P_{Ce}(\omega)$, defined by Eq. (D18), is the complex power associated with the electric current source; $P_{re}(\omega)$, the real part of $P_{Ce}(\omega)$, is seen from Eq. (D17) to be the time average of the real or irreversible power associated with the electric current sources. Similarly, $P_{ie}(\omega)$, the imaginary part of $P_{Ce}(\omega)$, is seen from the second term of Eq. (D17) to be the amplitude of the reversible or reactive power associated with the electric current sources.

The instantaneous power $P_{gm}(t)$ generated by the magnetic current sources only can be developed in a similar manner. As for the electric current density, the magnetic current density is also assumed to be of the form

$$\mathbf{J}_m(\mathbf{r}, \omega) = \mathbf{u}_m(\mathbf{r}, \omega) J_m(\mathbf{r}, \omega) \exp(i\theta_m) \tag{D19}$$

where $\mathbf{u}_m(\mathbf{r}, \omega)$ is a real unit vector and $J_m(\mathbf{r}, \omega)$ and θ_m are real numbers. Also, the magnetic current source is assumed to be localized, with θ_m being independent of $\mathbf{r}$. When the instantaneous powers due to both the electric and the magnetic current sources are included, Eq. (D17) is modified as

$$P_g(t) = P_{ge}(t) + P_{gm}(t) \tag{D20}$$

where $P_{ge}(t)$ is given by Eq. (D17):

$$P_{gm}(t) = P_{rm}(\omega) 2 \cos^2(\omega t - \theta_m) + P_{im}(\omega) \sin[2(\omega t - \theta_m)] \tag{D21}$$

and

$$P_{Cm}(\omega) = P_{rm}(\omega) + i P_{im}(\omega) = -\frac{c}{2} \int_{V_s} \mathbf{H}(\mathbf{r}, \omega) \cdot \mathbf{J}_m^*(\mathbf{r}, \omega) dV \tag{D22}$$

The interpretation of Eqs. (D21) and (D22) for the magnetic current sources is similar to that for Eqs. (D17) and (D18) for the electric current sources.

D.2 Vector potentials

From Eqs. (D8) and (D9), the electromagnetic fields are expressed in terms of the magnetic vector potential $\mathbf{A}$ and the electric vector potential $\mathbf{F}$ as [4,5]

$$\mathbf{E} = ik\mathbf{A} - \frac{1}{ik}\nabla\nabla \cdot \mathbf{A} - \nabla \times \mathbf{F} \tag{D23}$$

$$\mathbf{H} = \nabla \times \mathbf{A} + ik\mathbf{F} - \frac{1}{ik}\nabla\nabla \cdot \mathbf{F} \tag{D24}$$

where the wavenumber $k = \omega/c = \omega(\mu_0\varepsilon_0)^{1/2}$ and $\mathbf{A}$ and $\mathbf{F}$ satisfy the inhomogeneous wave equation

$$(\nabla^2 + k^2)(\mathbf{A}, \mathbf{F}) = (-\mathbf{J}_e, -\mathbf{J}_m) \tag{D25}$$

A reference direction is chosen to be along the x axis, with $\hat{x}$ being the unit vector along the x axis and the subscript t denoting directions transverse to the x axis. A transverse electric (TE) mode is constructed in terms of a single component $\hat{x}F_x$ of the electric vector potential $\mathbf{F}$ ($\mathbf{A}_t = 0$, $A_x = 0$ and $\mathbf{F}_t = 0$), yielding

$$E_x = 0 \tag{D26}$$

$$\mathbf{E}_t = \hat{x} \times \nabla_t F_x \tag{D27}$$

$$H_x = ikF_x - \frac{1}{ik}\frac{\partial^2 F_x}{\partial x^2} \tag{D28}$$

$$\mathbf{H}_t = -\frac{1}{ik}\nabla_t\frac{\partial F_x}{\partial x} \tag{D29}$$

where F_x satisfies the wave equation (D25). Similarly, a transverse magnetic (TM) mode is constructed in terms of a single component $\widehat{x}A_x$ of the magnetic vector potential $\mathbf{A}(\mathbf{A}_t = 0, \mathbf{F}_t = 0$ and $F_x = 0)$, yielding

$$E_x = ikA_x - \frac{1}{ik}\frac{\partial^2 A_x}{\partial x^2} \tag{D30}$$

$$\mathbf{E}_t = -\frac{1}{ik}\nabla_t\frac{\partial A_x}{\partial x} \tag{D31}$$

$$H_x = 0 \tag{D32}$$

$$\mathbf{H}_t = -\widehat{x} \times \nabla_t A_x \tag{D33}$$

where A_x satisfies the wave equation (D25).

References

1. S. R. Seshadri, *Fundamentals of Transmission Lines and Electromagnetic Fields* (Addison-Wesley, Reading, MA, 1971).

2. T. Padhi and S. R. Seshadri, "Radiated and reactive powers in a magnetoionic medium," *Proc. IEEE*, **56**, 1089–1090 (1968).

3. S. R. Seshadri, "Constituents of power of an electric dipole of finite size," *J. Opt. Soc. Am. A* **25**, 805–810 (2008).

4. J. A. Stratton, *Electromagnetic Theory* (McGraw-Hill, New York, 1941), Chaps. 1 and 6.

5. S. R. Seshadri, "Electromagnetic Gaussian beam," *J. Opt. Soc. Am. A* **15**, 2712–2719 (1998).

Airy integral

The Airy integral [1,2] is defined by the Fourier transform relations:

$$\text{Ai}(t) = \frac{1}{2\pi} \int_{-\infty}^{\infty} dq \exp(iqt) \exp\left(i\frac{q^3}{3}\right) \tag{E1}$$

$$\exp\left(i\frac{q^3}{3}\right) = \int_{-\infty}^{\infty} dt \exp(-iqt) \text{Ai}(t) \tag{E2}$$

In the investigation of Airy beams and waves, a general integral with a cubic phase term appears, as given by

$$I_A = \frac{1}{2\pi} \int_{-\infty}^{\infty} dq \exp\left(iqa_1 t + ia_3 \frac{q^3}{3} + ia_2 q^2 + b_1 q\right) \tag{E3}$$

The various coefficients should satisfy certain requirements that are generally met in physical problems. The q^3 and q^2 terms are written as part of a perfect cube as follows:

$$\begin{aligned}
E(q) &= ia_3 \frac{q^3}{3} + ia_2 q^2 + iq(a_1 t - ib_1) \\
&= i\frac{a_3}{3}\left(q + \frac{a_2}{a_3}\right)^3 + iq(a_1 t - ib_1 - a_2^2 a_3^{-1}) - \frac{i}{3} a_2^3 a_3^{-2}
\end{aligned} \tag{E4}$$

The variable of integration is changed from q to ξ, as defined by

$$\xi = a_3^{1/3}(q + a_2 a_3^{-1}) \quad \text{and} \quad d\xi = a_3^{1/3} dq \tag{E5}$$

By using Eq. (E5), $E(q)$ given by Eq. (E4) is expressed in terms of ξ as follows:

$$\begin{aligned}
E(q) &= \frac{i}{3}\xi^3 + i\xi(a_1 a_3^{-1/3} t - a_2^2 a_3^{-4/3} - ia_3^{-1/3} b_1) \\
&\quad - ia_1 a_2 a_3^{-1} t + \frac{2i}{3} a_2^3 a_3^{-2} - a_2 a_3^{-1} b_1
\end{aligned} \tag{E6}$$

Equations (E5) and (E6) are substituted into Eq. (E3); then, from the definition of the Airy integral given by Eq. (E1), I_A is determined as

$$I_A = a_3^{-1/3} \exp\left(-ia_1 a_2 a_3^{-1} t + \frac{2i}{3} a_2^3 a_3^{-2} - a_2 a_3^{-1} b_1 \right)$$

$$\times \mathrm{Ai}(a_1 a_3^{-1/3} t - a_2^2 a_3^{-4/3} - ia_3^{-1/3} b_1) \tag{E7}$$

References

1. M. Abramowitz and I. A. Stegun, *Handbook of Mathematical Functions* (Dover, New York, 1964), pp. 446–452.

2. J. C. P. Miller, *The Airy Integral, Mathematical Tables*, Part. Vol. B (Cambridge University Press, Cambridge, UK, 1946).

Index

γ_m 106, 119, 156
ε_m 101, 115, 156

Agrawal and Pattanayak framework 21–2
Airy beams 181
  fundamental 181–4
  modified fundamental 184–7
Airy function 182
  Fourier transform 183, 185
Airy integral 219–20
Airy wave 181
  basic full modified 190–7
  fundamental 187–90
analytical continuation 36, 48–50
asymptotic Gaussian beam field 47–8
atoms, guiding and focusing 82
azimuthal mode number 98, 102, 111, 112, 152
azimuthal polarization 68

basic full cosh-Gauss wave 93–4
basic full Gaussian wave 35–43, 196, 197
  complex space, point source in 35–40
  radiation intensity 40–1
  radiative powers 41–2
  reactive powers 41–2
  time-averaged power 37–9
basic full real-argument Laguerre–Gauss wave
  111
beams 1
beam shape parameter 152
beam spreading 7–8
beam waist 28, 152
beam waveguide 9
beam width 6, 39, 61
  of fundamental Gaussian wave 18

Bessel function 69
Bessel integral 100, 103
Bessel transform 74, 86, 98, 112
  of Green's functions 209–11
  pair 69, 72, 203
  relations 99
  of scalar Green's function 203–4
branch cut 30, 53
  choice of 53
branch line 29
branch point 51, 52
  analytical continuation 52

coherence length 167
complex-argument Hermite–Gauss beam 123
  paraxial beam 123–6
  time-averaged power 126–8
complex-argument Hermite–Gauss wave 123,
  128–30
  real and reactive powers 130–4
complex-argument Laguerre–Gauss beam 97
  paraxial beam 97–100
  secondary source plane 98
  time-averaged power 100–1
complex-argument Laguerre–Gauss wave 103–4
complex distance 51
complex number 27
complex power 6–7, 13, 19–20, 41, 55, 63–4,
  100, 105–7, 114–16, 119, 142, 156, 189,
  195–6, 215
complex rays 33, 42
complex space 55–6, 73
  current source in 72–4
  current source of finite extent in 55–59
  four point sources 92–3

higher-order point source 87–8
point source 45–7
complex space, point current source in 25
exact solution 30
extensions 29–30
field of point source 28–9
scalar Gaussian beam 26–8
complex space, point source in
basic full Gaussian wave 35–40
complex source point theory 45
analytic continuation 48–50
asymptotic field 47–8
derivation of complex space source 45–7
limiting absorption 50–3
complex space source 55–7, 87–8, 92–3,
101–3, 128–9, 133, 140–1, 148–9, 155,
163, 192
derivation 45–7
higher order point sources 116–17
theory 161
confocal ellipses and hyperbolas 30
cosh-Gauss beam 90–2
cosh-Gauss wave 93–4
cross-spectral density 162, 165–9
Cullen and Yu 43
current density 68
electric 6
magnetic 8–9
current source of finite extent 56
cylindrical coordinate system 50, 97
cylindrically symmetrical beam 67
cylindrically symmetric complex-argument
Laguerre–Gauss beam
full-wave generalization of 102
cylindrically symmetric TM beam 67–72
cylindrically symmetric TM full Gaussian wave
67, 74–5
current source in complex space 72–4
cylindrically symmetric TM full wave 74–5
full-wave generalization of 67
radiation intensity distribution 80–2
reactive power 78–80
real power 75–8

delta function 199
Deschamps 25, 29
Deschamps' *Letter* 31–4
Deschamps' novel virtual source 25, 28–9
detector response time 166

difference coordinate 169
duality 68

electric current 111
electric current density 13, 100, 105, 114, 118,
126, 131, 184, 187, 195
electromagnetic fields 67, 100, 126, 165, 184,
189, 213
components 37
paraxial beam 3–4
Poynting vector and generated power 213–16
vector potentials 216–17
E-plane 4, 176
evanescent wave 2
exact electromagnetic fields 15
of Gaussian wave 15–17
exact solution 30
exact vector potential 13–14
extended full Gaussian wave 55
of Gaussian wave 13–14
excitation coefficient 36, 45, 48, 56, 58, 73, 75,
89, 104, 118, 129, 130, 163, 164, 192
exponential integral 205
extended full Gaussian wave 55–66, 161, 162–5
complex space, current source of 55–8
exact vector potential 55
radiation intensity 61–3
radiative powers 63–6
reactive powers 63–6
time-averaged power 55, 58–61
extended full-wave generalization of paraxial
beam 162–5
extensions 29–30

f_0 4
Fabry-Perot interferometer 10
far-field optics 10
field of point source 28–9
finite dimensions of source 42
finite-energy fundamental Airy beam 181
fluctuation parameter 170, 172
fluctuation period 161, 165
focal plane 152
Fourier transform pair
three-dimensional 199
two-dimensional 201–2
free-space electromagnetic wave velocity 46
Fresnel integral representation 27
fully incoherent source 168

fundamental Airy beam 181–4
  modified 184–7
fundamental Airy wave 181, 187–90
  modified 190–7
fundamental Gaussian beam 13, 35, 67, 94,
    97, 111
  full-wave generalization of 45
fundamental Gaussian wave 13–23, 67

Gamma function 127
Gaussian beam 1
  applications 9–10
  as complex rays 31
  beyond paraxial approximation 21–2
  electromagnetic fields 3–5
  fundamental 1–11
  limitations 9–10
  magnetic current density 8–9
  radiation intensity 4–6
  radiative powers 6
  reactive powers 7
  spreading on propagation 7–8
  vector potential 1–3
Gaussian profile 1
Gaussian wave
  exact electromagnetic fields 15–17
  exact vector potential 13–14
  fundamental 13–23
  Gaussian beam
    beyond paraxial approximation 21–2
  radiation intensity 17–19
  radiative powers 19–21
  reactive powers 19–21
generalized Laguerre polynomial 98, 112
  orthogonality relation 115
  series expansion 116
generated power 214
Goubau beam waveguide 31
Green's function 29
  Bessel transforms of 209–11
  Fourier transform of 209
  for Helmholtz equation 199–204
  for paraxial equation 207–11
guiding and focusing of atoms 82

Helmholtz equation 2, 22, 26, 36, 86, 129, 141,
    188, 193
  Green's function for 199–204
  reduced 103, 112, 155

Helmholtz wave equation 13, 45, 56, 73
Hermite polynomial 124
  orthogonality relation 140
higher-order complex space source 101–3
higher-order Gaussian beams 97, 111
higher-order hollow full Gaussian wave 89
higher-order hollow Gaussian beam 85–7
higher-order hollow Gaussian wave 85
  complex space source 87–8
  hollow Gaussian wave 88–9
  paraxial beam 85–7
higher-order point source 72–3
higher-order source 73–4
hollow Gaussian wave 88–9
hollow waves 82
H-plane 4, 177

input properties 29
integral 205
inverse Bessel transform 75
inverse Fourier transform 3, 13–14, 58, 91

Laguerre polynomial 98, 115
Landesman and Barrett, exact solution by 30
length parameter 55, 67
limiting absorption 50–3
location coordinates 45
lowest-order solution 67

magnetic current density 72
magnetic vector potential 98, 104, 112, 114, 118,
    123, 126, 139, 151, 153, 156, 165, 183, 187,
    188, 194
Maxwell's equations 161
mean-squared width 56
mode number 98, 112
mode numbers m, n 123
mode numbers n, m 97, 111
  lowest order 97
mode numbers $n = 0, m$ 97, 111
mode numbers $n = 0, m = 0$ 97, 111, 123, 137
modified Bessel–Gauss beam 151–3
modified Bessel–Gauss wave 151, 155–6
  expression for the complex power of
    156
  paraxial beam 151–3
  real and reactive powers 156–8
  time-averaged power 153–5
modified Schell model source 162

near-field optics 10
normalization constant 46, 71, 98, 101, 112, 115, 140
normalized Poynting vector 8
normalized width 6

observation point 52, 53
optical system 32
orbital angular momentum 82

paraxial approximation 36, 207–8
paraxial beam 1, 85–7, 123–6, 137–9, 161
    extended full-wave generalization of 162–5
paraxial electromagnetic Gaussian beam 29
paraxial wave equation 1, 2, 26, 56, 69
partially coherent full Gaussian wave 161
    radiation intensity for 169–73
    time-averaged power for 173–5
partially coherent source 167–8, 171
partially incoherent full Gaussian wave 161
    radiation intensity for 175–8
    time-averaged power for 178
partially incoherent source 168–9, 171
physical spaces 51
plane circular sheet 29–30
plane-wave phase 56
plane-wave phase factor 69
point current source, in complex space 25
    exact solution 30
    extensions 29–30
    field of point source 28–9
    scalar Gaussian beam 26–8
point electric dipole 25
point source in complex space 35
polarization 81–2
power
    complex 6–7
    reactive 6–7
    real 6–7
Poynting vector 7, 162, 213–14
    and generated power 213–16
    time-averaged 166

quality of paraxial beam 17, 39, 60, 81

radial mode number 98, 111, 112
radial polarization 68
radiation fields 29

radiation intensity 4–6, 17–19, 39–40, 61–3, 71, 80–2, 169–73, 175–8, 190, 196–7
    basic full Gaussian wave 39–40
    extended full Gaussian wave 61–3
    of Gaussian wave 17–19
    partially coherent source 172
    partially incoherent source 176
radiative power 6–7, 17, 20, 41, 63–4
radius of curvature 27
rapidly varying phase 1, 2, 26, 69, 98
rate of spreading 56
Rayleigh distance 2, 28, 67, 69, 86, 162
    equivalent 193
reactive power 13, 20–1, 29, 42, 55, 64–6, 78–80, 100, 107, 114, 118–22, 127, 132–3, 137, 139, 142–3, 153, 157, 184, 187, 190
real-argument Hermite–Gauss beam 137
    paraxial beam 137–9
    time-averaged power 139–40
real-argument Hermite–Gauss wave 137, 140–1
    complex space source 148–9
    evaluation of integral $I_m(x,z)$ 147–8
    real and reactive powers 142–7
real-argument Laguerre–Gauss beam
    complex power 114–16
    paraxial beam 111–14
real-argument Laguerre-Gauss wave 116–18
real basic Rayleigh distance 50
real dipole 26
real power 1, 6–7, 20, 41, 64, 75–8, 104–8, 115, 118–22, 128, 133–4, 137, 140, 143–7, 153–4, 157–8, 184, 187, 190
    for basic full Gaussian wave 41
    cylindrically symmetric transverse magnetic full Gaussian wave 75–8
rectangular coordinate system 50
resonance 21, 65
resonator modes 29

scalar Gaussian beam 26–8
scalar Green's function
    Bessel transform 203–4
    Fourier transform 201–2
    three-dimensional 199–201
scalar point source 35, 45
Schell-model source 167–8
    modified 168

slowly varying amplitude 1, 2, 26, 69
Sommerfeld's formula 204
source electric current density 6, 45, 55
source magnetic current density 9
source of finite dimensions 29
source surface electric current 111
spherical coordinates 4, 5, 71
spreading of beam 27
strength of source 37
sum coordinate 169
surface electric current density 40, 45, 55, 63

TE Gaussian wave 81–2
time-averaged power 13, 16–17, 38, 41, 58–61,
   71, 77–8, 100–1, 153–5, 173–5
   complex-argument Hermite–Gauss beam
      126–8
   fully incoherent source 175
   partially coherent source 172
   partially incoherent source 176, 178–9
   real-argument Hermite–Gauss beam
      139–40
time-averaged Poynting vector 4, 15, 70,
   166
time-averaged value 214–15
TM full Gaussian wave 74–5
TM Gaussian wave 81–2
transverse electric (TE) beam 67
transverse electric (TE) waves 67
transverse Laplacian operator 101
transverse magnetic (TM) beam 67
transverse magnetic (TM) mode 217

transverse magnetic (TM) modified Bessel–Gauss
   beam 151
transverse magnetic (TM) waves 67
trigonometric terms 106
two-dimensional Fourier transform 2
   pair 201–2
two-dimensional inverse Fourier transform 58
two higher-order full Gaussian waves 85

Union Radio-Scientifique Internationale
   (URSI) 28

variance 27
vector potential
   electric 1, 8, 216
   magnetic 1, 216–17
virtual dipole 26
virtual point electric dipole 25
virtual point source 25
   basic full Gaussian wave 47
   paraxial beam 46

waist 162
   equivalent 192
   size 2, 10
wavenumber 2, 51, 98, 112
wave period 161
wave velocity
   free space 2

zero order Bessel function 87–8

Printed in the USA
CPSIA information can be obtained
at www.ICGtesting.com
JSHW051410221024
72173JS00006B/1332